Managing the Mean Math Blues

Math Study Skills for Student Success

SECOND EDITION

Cheryl Ooten
Santa Ana College

with

Kathy Moore
Santiago Canyon College

Boston   Columbus   Indianapolis   New York   San Francisco   Upper Saddle River
Amsterdam   Cape Town   Dubai   London   Madrid   Milan   Munich   Paris   Montreal   Toronto
Delhi   Mexico City   Sao Paulo   Sydney   Hong Kong   Seoul   Singapore   Taipei   Tokyo

Vice President and Editor in Chief: Jeffery W. Johnston
Executive Editor: Sande Johnson
Editorial Assistant: Lynda Cramer
Vice President, Director of Marketing and Sales Strategies: Emily Williams Knight
Vice President, Director of Marketing: Quinn Perkson
Executive Marketing Manager: Amy Judd
Senior Managing Editor: Pamela D. Bennett
Project Manager: Kerry J. Rudadue

Senior Operations Supervisor: Matthew Ottenweller
Art Director: Candace Rowley
Cover Designer: Candace Rowley
Full-Service Project Management: Thistle Hill Publishing Services, LLC
Composition: Integra Software Services
Printer/Binder: Bind-Rite Graphics
Cover Printer: Bind-Rite Graphics
Text Font: Berkeley

Credits and acknowledgments borrowed from other sources and reproduced, with permission, in this textbook appear on appropriate page within text.

Every effort has been made to provide accurate and current Internet information in this book. However, the Internet and information posted on it are constantly changing, so it is inevitable that some of the Internet addresses listed in this textbook will change.

Library of Congress Cataloging-in-Publication Data

Ooten, Cheryl.
    Managing the mean math blues : math study skills for student success/Cheryl Ooten
with Kathy Moore.—2nd ed.
        p. cm.
    Includes bibliographical references and index.
    ISBN-13: 978-0-13-229515-4 (pbk.)
    ISBN-10: 0-13-229515-6 (pbk.)
    1. Mathematics—Study and teaching—Psychological aspects.   I. Moore, Kathy.
II. Title.
QA11.2.O67 2010
510.71—dc22

                                                            2009019000

10 9 8 7 6 5 4 3 2 1

www.pearsonhighered.com

ISBN 10:   0-13-229515-6
ISBN 13: 978-0-13-229515-4

To my husband, Bob, for his continued support; to my parents, R. J. and Doris Thomas, for modeling curiosity; and to my students, who kept coming to class.

—Cheryl Ooten

To my husband, Mike, and my parents, Jan and Dick—with all my love—thank you for everything.

—Kathy Moore

BRIEF CONTENTS

CONTENTS

PREFACE

Math courses are "gatekeeper" courses to many well-paid and respected professions, and the gate continues to slam in the faces of promising students of all ethnicities, genders, and ages. This book addresses the issues that stop those students. Those issues are misconceptions about math and who can learn it, lack of study skills, and math anxiety. In the first-person voice of the authors (both experienced math teachers) and the voices and stories of seven successful math students, this book guides the reader to self-efficacy and success with math.

Designed for the adult student who did not complete the math series in high school and who needs an alternative approach, the second edition of *Managing the Mean Math Blues* will help students act directly, set doable short-term goals, incorporate brain-based learning techniques, review basic math concepts, lower their anxiety, and *finally* know how to study math.

With an understanding of reframing negativity about math, students will examine their belief systems and their unique learning styles, moving on to practice math study skills, test-taking strategies, problem solving, and managing math anxiety. Each chapter ends with activities to solidify the chapter content by practicing math. The book ends with an optional overview of basic algebra to give students a framework on which to build their continued math development.

The second edition of *Managing the Mean Math Blues: Math Study Skills for Student Success* is especially suited as an ongoing supplement to a regular basic math or beginning algebra course as well as separate courses on study skills or overcoming math anxiety. This book's chapters stand alone so that classroom issues can be addressed as the instructor sees fit. Three chapters of math study skills have been added, and all of the first edition chapters have been revised and placed in a useful sequence. Also, successful students' stories, reviews of basic skills, follow-up activity questions, and an optional algebra chapter give the instructor many options for meeting the needs of their diverse students.

The *Online Instructor's Manual* for *Managing the Mean Math Blues: Math Study Skills for Student Success* includes teaching suggestions, worksheets, and classroom activities to give students hands-on practice developing new skills.

Acknowledgments

Cheryl Ooten

Writing a book is both a solitary and a community project. Although the act of writing itself is done alone, the ideas and inspiration come from others. To all of those people, I express my gratitude. Some of them, such as my editor, Sande Johnson; my reviewers; my new co-author, Kathy Moore; my co-illustrator, Emily Meek; my husband, Bob Ooten; and my students have been constant and vigilant in the process. The others have been no less important in the final development of this book, and I thank them from the bottom of my heart. Special thanks go to Mihaly Csikszentmihalyi for his inspiration and support. Also, I appreciate the contributions of, in alphabetical order, Ruth Afflack, Dennis Airey, Mary Anne Anthony, Elinor Peace Bailey, Phyllis Biel, Cherie Bowers, Bill Browne, Linda Browne, Tim Cooley, Lynda Cramer, Enrique De Leon, David Drew, Melba Finklestein, Dennis Gilmour, Mary Halvorson, Anne Hauscarriague, Jazmin Hurtado, Sarah Kershaw, Suzi Lohmann, Lynn Marecek, Don McIntyre, Russ Meek, MESA students at Santa Ana College, Tom Mock, Yolanda Mugica, Bobbi Nesheim, Carlene Ono and her Writing Group, Blanche and LaVerne Ooten, Carlos Ordiano, Claire Parker, Craig Parker, Josh Parker, Kati Ooten Parker, Cerise Paton, Maureen Pelling and her Writing Group, Mike Petyo, Giovanna Piazza, Jack Porter, Tony Rivas, Kerry Rubadue, Judy Schaftenaar, Joel Sheldon, Alex Solano, Christa Solheid, Virginia Starrett, Kathy Taylor, Adam Thomas, R. J. and Doris Thomas, Carol Tipper, Angela Urquhart, Melody Vaught, Isabella Vescey, Bob White, David Wintle, Karin Wright, and my wonderful math department and other colleagues at Santa Ana College.

I also wish to thank the many reviewers who have diligently read chapters and given valuable feedback both for the first edition and for this new edition. For this edition: Nancy D. Eschen, Florida Community College at Jacksonville; Louise Hillery, Ivy Tech in Columbus, Indiana; Hubert J. Horan, University of New Mexico; Linda Mudge, Illinois Valley Community College; Elsie Newman, Owens Community College; Jennifer Sawyer, Currituck County Schools; April Strom, Scottsdale Community College; Richard H. Sturgeon, University of Southern Maine; Joan Totten, Ferris State University; and Justine Wong, Serra Preschool and National Hispanic University.

Kathy Moore

I would like to thank:

- all my students for working so hard and giving me the insight into their world.
- Anne, Laney, Joyce, Kim, and Amy for the road trips.
- all my co-workers at Santiago Canyon College for their support and sharing of ideas.
- My family and friends, including the canine variety, for being understanding and patient.
- Cheryl for giving me the chance to go on such a wonderful adventure in learning.

Limits
Limits
Limits
Limits
Limits

I N T R O D U C T I O N

"My favorite thing is to go where I've never been."

–DIANE ARBUS

We all have limitations. We often think our limitations are different and larger than they truly are. I believe that about you and math. To identify and push past limitations that you have set for yourself, this book will teach you how to take charge of math and to bring any difficulties under your control so you can push your math skills further than you now believe possible.

As a math teacher, I have been told by countless students what they believed were their limitations:

■ "In fourth grade, my teacher began fractions. I was clueless. When I see a fraction now, my mind closes."

■ "I loved math in elementary school. As I started algebra, everything changed. Math made no sense anymore."

■ "I do my homework and I understand my homework but in math exams, I forget everything."

■ "In class, I follow the teacher. At homework time, the problems are beyond me."

■ "All my life I've dreamed of being a teacher. The problem is that Liberal Arts Math is required and I have not passed the prerequisites."

The limitations these students describe can be changed with new perspectives and different study skills.

Some of these students experience what I call the "Mean Math Blues" or what some people call "math anxiety."

Signs of the Mean Math Blues			
anger	boredom	frustration	avoidance
sadness	confusion	fear	helplessness
nausea	tension	headaches	palpitations

You may have thought some of the same things my students tell me or you might have experienced a few of the signs.

You Are Not Alone!

More than one third of my math students share similar feelings and experiences. In survey after survey from basic math through calculus, students tell me they experience anxiety about math or about math tests. As a math teacher, this concerns and saddens me. I know my math colleagues feel the same. We all love math and find it, for the most part, exciting and challenging. We would like to share with you the experience of finding joy in math.

I Got the Blues

What might make me different from many of my math colleagues is that during my second year of graduate school in math, I experienced math anxiety or what I call the Mean Math Blues. Each Tuesday and Thursday morning of that fall semester, I woke with a feeling of dread as I realized I had to go to class. I sat in front, desperately trying to write everything said or written on the board. My brain felt like cotton. The professor's voice came from far away. I thought myself too shy to ask fellow students questions and for assistance with homework.

As the semester dragged on, I felt more and more relief as class was dismissed. I avoided homework until, you guessed it, right before the next class when my high anxiety level prevented clear thinking. I lost confidence and endured a miserable semester.

How Do You Spell Relief?

When the title of Sheila Tobias's book *Overcoming Math Anxiety* jumped off the shelf in the Glendale Galleria bookstore and filled me with relief, I found words of explanation for my difficult experience. Those two little words—math anxiety—helped me separate myself from my experience, give it a name, and take control over it. "Math anxiety" reframed what had happened, giving me a new perspective. I began reading everything I could find about anxiety and math—discovering many ways I could have thought and behaved differently in my math course to change the experience. It is those revelations and my experiences with scores of math students since that I share with you in this book.

Reframing Can Change the Experience

Reframing is a powerful strategy for changing our perceptions. It puts a different context or frame around an event or experience. A reframe you might already know is to change the thought "The glass is half empty" to "The glass is half full." Both statements are true, but one focuses on what's missing while the other focuses on what exists.

In a reframe, both thoughts are possible but one thought has a narrow focus while the other opens to a larger perspective. A reframe can change our view of an experience and allow us to see increased and sometimes inspired options for action. This book is full of reframes that will enable you to increase your math power. If you just change "I don't understand" to "I don't understand *yet*," you open up the possibilities for your understanding in the future.

"Yet" is a powerful word that can reframe many math thoughts such as "I can't do this," "I'm not done with my homework," or "I'm not prepared for my test." Reread these three sentences, putting "yet" at the end of each. Notice how this changes the meaning.

This humorous, nonmath example of reframing might clarify the power of reframing even further. Suppose you are driving three children under the age of 10 to visit their grandmother, a three hour drive away. If you think that there will be no fights in the backseat, you set yourself up for disappointment as well as for a very long drive. If, however, you predict that there will be at least one fight every half hour and it turns out that there is only one fight every hour, you have reframed your expectations to make the trip more pleasant for all.

Reframing your expectations for your math studies can make your trip through math more pleasant. For example, if you think that you should understand math immediately or should do all math problems quickly, you set yourself up for frequent disappointment. However, if you recognize that comprehending math takes time, repetition, and making mistakes, you have reframed your expectations to create a more positive, proactive experience.

You Can Make New Connections

Recent brain research has discovered that human brains have limitless potential for learning at all ages. These discoveries promise new possibilities for math students who are willing to push into new territory. This book will support you in many ways to do "something new and different" as you work with math. This book can help you learn new math strategies, mobilize your support resources, and take charge of your thoughts and actions in powerful ways.

You can change avoidance in math to excitement about conquering challenges. Conquering the math you need is a process—a process that takes intention and energy. *It will not happen overnight.* Sometimes you will progress quickly. Sometimes, slowly. As you improve your study skills, you will have more energy to experiment with new successful learning strategies.

How to Use This Book

Because every reader has different wants and needs, these chapters stand alone and can be read in any order. Some readers will be in a math class, while others will be thinking about taking a math course. Don't expect to use every suggested strategy. A few good ideas that work for you can make the difference between success and failure. Experiment with different ideas to find what works best for you. Keep this book as a reference tool to serve your changing needs as you study and use math. **See the Problem-Solving Index in the Appendix to find specific solutions for individual problems.**

CHAPTER 1

Begin with Facts

"Give me where to stand and I will move the earth."

ARCHIMEDES

Are You Cutting Yourself Short?

A young woman questioned her mother, who was preparing the traditional roast for the family holiday celebration. Every year the mother cut off both ends of the roast and put all the pieces into a pan in the oven. "Why do you cut off the ends?" the young woman asked. The mother responded that *her* mother had always done that. When the young woman questioned her grandmother, the grandmother said she cut off the ends because she had never owned a pan large enough to hold an entire roast.

Limiting beliefs unconsciously pass down through cultures—the cultures of families, ethnic groups, religious groups, and institutions. Unexamined beliefs may cause you to do something just because your grandmother or society did not have the resources or information to do differently. **Even your math work may be influenced by the unconscious, limiting beliefs you have absorbed** (Kogelman & Warren, 1978). To check your beliefs about math, answer the questions in the quiz.

QUIZ

Answer True or False before reading on to learn about your math beliefs.

_____   **1.** Mathematicians do math quickly in their heads.

_____   **2.** I can't do math.

_____   **3.** Math is always hard.

_____   **4.** Only smart people can do math.

_____   **5.** If I don't understand a problem immediately, I never will.

_____   **6.** There is only one right way to work a math problem.

_____   **7.** I am too shy to ask questions.

_____   **8.** It is bad to count on my fingers.

_____   **9.** To complete my math requirements quickly, I should skip to the highest level math class that I can.

_____   **10.** My negative math memories will never go away.

Each statement on this quiz is *false*. **Are you surprised?** These misconceptions are negative math beliefs acquired subtly through cultures, families, teachers, and friends. For math students, negative beliefs can be deadly. At the very core of your self-esteem, these negative beliefs keep you uneasy.

Belief in these misconceptions is a stumbling block to learning math because you think and act from your belief system. The following sections explain why I believe the quiz statements are all false and how I reframe them. Notice which statements you marked True. Read those sections first.

Challenge Your Negative Beliefs

1. Mathematicians Do Math Quickly in Their Heads

The only problems mathematicians complete quickly in their heads are the ones they've done many times before. There are math problems that have remained unsolved for hundreds of years, and no speedy mathematician has ever solved them.

How do mathematicians work math at their own level? At their own level—a level that challenges them, and where they can advance—they use scratch paper—usually lots of it. They write down the problem, the ideas in the problem, what they are trying to do, any thoughts they have, possible solutions, and so on. They make many guesses and check out those guesses. They draw pictures and diagrams. They look up definitions of words in the problem. They talk to others. They start at the end of the problem and try to work backward. They begin again on clean paper to think in new ways. They know that working at this level takes patience and time. **They don't give up and are willing to begin again and again** until they achieve success. In Chapter 9, you will learn problem-solving strategies that math-ematicians use frequently.

Brain research supports the idea that speed in mathematics comes from practice, which solidifies the connections and pathways in the brain. The key to speed in math is to slow down first, understand the material, and then practice.

Reframe your math work. Change from saying "Mathematicians do math quickly in their heads" to saying "Mathematicians can quickly do problems that they have practiced. When they learn new math ideas, they work slowly and often begin again."

2. I Can't Do Math

Not true! If you can count, add numbers, recognize circular and rectangular shapes, and point to the front and the back of the classroom, you can do math. All these skills are math. You do math all the time. Knowing your age, comparing sizes and shapes, adding your money, and subtracting to get change are math skills.

You use math every day of your life at home or at work without giving it a second thought. When you drive, you judge distances, speeds, and times. You estimate if you can afford a vaca-tion or a car. You compare volumes as you cook and areas as you rearrange furniture in your home. You use statistics as you watch sports and consider things like RBIs in baseball or field goal percentages in basketball. All of these are mathematical skills taken for granted.

Sometimes students are placed in math classes beyond their skill level and become discouraged. Other students enroll in the appropriate course but don't do the work to keep their skills matching the challenge of the course, and they become discouraged. Discouraged students often generalize and say, "I can't do math." **The way to regain your confidence is to slow down and discover the level of math where you are challenged and learn new concepts, but are not overwhelmed. Then set up all the time and support that will keep you learning.**

Reframe your math work. Change from saying "I can't do math" to saying "I can already do some math and I can learn more."

3. Math Is Always Hard

There are some math concepts that are hard for everyone. There are math problems that have been unsolved for hundreds of years even though they've been attempted by competent, knowledgeable mathematicians who work at them for decades. Those aren't the problems you need to work, unless you are curious. When you work at your appropriate level, you find a combination of easy ideas and hard ideas. If all of the ideas are hard, you may need to seek more support or to spend more time studying in a different way.

You may get discouraged when you compare your speed and understanding in math with your teacher's speed and understanding. Math teachers appear to do math easily in class because they have done these specific problems, or ones like them, many, many times. Also, the problems you see them work are not at the "edge" of their skill level.

You will want to study and progress at your own "growing edge"—the skill level where you have a bit of discomfort with new material but where you are not totally overwhelmed. This is the place where your skills match the math challenges. You can expect challenges that may trouble you at first, but they can be overcome. Credit yourself often by noticing how much more math you know now than you did two weeks ago. We often discredit the math we do easily. And we often berate ourselves for struggling with concepts such as negative numbers. Reportedly, "some of the best mathematicians in history shared those same struggles and frustrations" about negative numbers (Berlinghoff, p. 98).

If you persist, reread your textbook and notes, and ask lots of questions, you can work through problems that are difficult in the beginning. Everyone is different and finds some problems easy and some problems hard. Finally understanding something you've struggled over can be satisfying. **Paying attention to details and reviewing to find what you've missed, as well as asking for help when you need it, are keys to success in math.**

Reframe your math work. Change from saying "Math is always hard" to saying "There are both easy and hard math problems for everyone. I will keep a match between my skills and the challenges."

4. Only Smart People Can Do Math

Being "smart" (whatever that means) could be helpful in math *but* the most important thing is to *keep working. Persistent students are successful.* They come to class. They read the textbook. They ask questions. They take notes. They ask questions. They do their homework. They ask questions.

I've had many so-called smart students who don't do these things. Because they think they know it all, they miss concepts, don't ask questions, become overwhelmed, and drop before the end of class. Often, as class begins, they act as if they already know everything and somehow seem unable to give up that role. Then when they don't understand, they are probably embarrassed to ask questions. Don't be influenced or intimidated by these students in your classes. Be persistent regardless of how everyone else acts. You will succeed. In Chapter 3, you will learn about different ways of being smart.

My student Michelle ignored all of the others in class as she sat in the front row of my geometry class asking question after question. Sometimes students sitting behind her snickered and laughed because her questions were so intense and the answers seemed so obvious. She was so concentrated on her own learning that she didn't even appear to notice the others. I admit that as Michelle's teacher, I had concerns about her future in math until three semesters later when I saw her in the hallway carrying a calculus book. She reported that she had passed class after class of math requirements and was now completing her final semester.

Michelle taught me a great lesson with her focused style. As a teacher, I learned I could not predict success for my students based on what I saw in class. I learned that persistence counts more, and I wondered how far my snickering students with the higher grades had come.

Reframe your math work. Change from saying "Only smart people can do math" to saying "Persistence in math helps more than smarts. Discovering how I am smart can help me use my strengths to learn math my way."

5. If I Don't Understand Immediately, I Never Will

Not true! When you do not understand something, you can say to yourself, "I do not understand *yet.*" The word "yet" opens the doors in your mind for understanding to still occur.

Productive people know that understanding follows the groundwork of reading, working problems, asking questions, thinking, reworking, and then relaxing for a while. How often have you tried and tried to solve a problem and then, after you finally gave up, had the solution pop into your mind?

One of the most important study skills that you can learn is to expect to *not* understand *immediately*! Know that understanding follows working, practicing, reading, asking questions, and living with the ideas for a while.

Some math concepts can be understood immediately. Some of them take mulling over and working with for a time—sometimes for a *long* time. Mathematical historian Lancelot Hogben said instructors often invite their students to quickly solve math dilemmas that took centuries or more for the human mind to clarify and understand. I can think of math concepts that I finally really understood only as I was discussing them in my classroom as the teacher. Patience with yourself is important.

Reframe your math work. Change from saying "If I don't understand immediately, I never will" to saying "Understanding math takes time. I don't understand this yet."

6. There Is Only One Right Way to Work a Problem

No, there are many right ways to do every math problem that exists. Actually, the "right way" to first work a problem is the way you understand it best. As you increase your understanding, you can be more open to other ways of looking at it. Having an open mind that is flexible and willing to experiment with different methods will increase your problem-solving skills.

When you are learning in a math class, your instructor may request that you work certain problems in specific ways because she knows that those methods will assist your understanding later on or make you more accurate. Remind yourself that your instructor can see the overall view from a broader perspective.

Reframe your math work. Change from saying "There is only one right way to work a problem" to saying "There are many right ways to work and think about each math problem."

7. I Am Too Shy to Ask Questions

Brains and bodies are dynamic. They change constantly and we have influence over those changes, so be careful about making this generalization about yourself. "*Feeling* shy" is different from "*being* shy." *Feeling* shy is a common experience when we find ourselves in a new or uncomfortable situation. *Being* shy is a label that many of us give ourselves, which causes us to choose shy behaviors. Your thoughts and your behaviors are the most powerful change agents you own, and they affect your emotions and body sensations. See Chapter 10 to learn more about these interactions. **As you practice talking to people and asking questions, choosing safe and supportive places to do that, you will notice your comfort and confidence levels rising.**

Believe it or not, the rest of the world has better things to do than watch us all the time. Notice that most people are more concerned with themselves than with you. They wonder how *you* see *them*. They busy themselves thinking what they will say next. Fortunately, or unfortunately, you just aren't so important to others that you become their focus of attention for more than a short time.

Often students find that speaking one on one with the instructor or another student increases their confidence in class. Even asking a question in class to which you already know the answer may increase your confidence. Asking your question before class or putting problems on the board before class will help too.

Colleges and universities have counseling assistance for students. A good counselor can assist you in finding safe campus resources for getting your math questions answered if feeling shy is a severe and disabling problem.

Reframe your math work. Change from saying "I am too shy to ask questions" to saying "Feeling shy or reluctant to ask questions is common. I can seek out safe situations to ask what I need to know. The more I practice asking, the more comfortable I will feel asking."

8. It Is Bad to Count on My Fingers

Not so! Our number system is based on 10—probably because we have 10 fingers and 10 toes. Consequently, we have a portable model of our number system with us at all times.

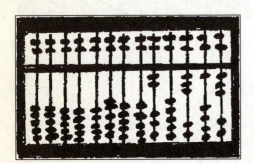

When children first learn about numbers, they need concrete models to understand what the abstract ideas mean. Unknowing teachers sometimes skip this concrete model stage.

Cultures in Asia, Europe, and Africa have used ingenious forms of the abacus to do complicated number calculations speedily.

Using sand, sticks, wires, rocks, and beads in various ways, these people comfortably worked with numbers in a concrete, representational way. If we need to, we can do the same with our fingers.

Many children and adults process information best through their skin and muscles by touching, feeling, and doing rather than by seeing or hearing. These people, called kinesthetic learners, particularly need the concrete connection with numbers that their fingers provide.

Using fingers freely in calculations, students become more confident and faster in their use. Eventually, most students move beyond this use unless they are in a new or stressful situation in which they need grounding. Continued practice with basic math facts helps us compute more quickly and accurately.

I teach students how to count on their fingers. Try it. Hold out your hands, palms up, so that your fingers are separated and not touching anything. Imagine an electrical charge running through your fingers. Feel numbers such as 3, 8, 8 + 3, 11, and so on. The more you practice simple problems like these, the faster you will become. You may find that you don't even need to move your fingers in order to count. No one else needs to know what you are doing.

I myself, a math teacher, have used my fingers to keep focused as I calculate money in a distracting environment. I have counted on my fingers as a volunteer cashier in a snack booth during a professional football game. I also use my fingers to count rests as I play the piano in a jazz band. Notice how musicians use their feet, hands, or heads to keep track of time as they perform. Would we dare tell them not to? Unfortunately, in math, using fingers to assist calculations has often been forbidden or ridiculed.

Reframe your math work. Change from saying "It is bad to count on my fingers" to saying "Using my fingers (if I need them) to calculate can be quick and helpful. Practice increases speed."

9. I Should Skip to the Highest Level Math Class

Placing yourself in a math class beyond your current skill level can cause the Mean Math Blues. In order to maintain math skills, you must practice them. When you are not using specific math skills, you forget them. They are not gone forever from your brain, but the connections (called dendrites) that your brain grows to form ideas weaken without use. To reactivate these skills, you need to relearn and practice them. Students who need to extend their math knowledge often find the fastest way is by repeating previous coursework before continuing. Success with math requires a constant match between your current skills and the challenges of the course.

"Vitality shows not only in the ability to persist but in the ability to start over."

F. Scott Fitzgerald

My personal experience is that these skills often return with more clarity and understanding the second time around. Brain research supports my experience, showing that the relearning process actually forms many new connections in the brain as well as reinforces the old ones.

Because math skills depend on facility with prerequisite skills, in order to advance it may be necessary to repeat a math class. There is no shame in forgetting math skills. It is the biological human condition to forget the skills you are not currently using. **Brain researchers say**, "**Use it or lose it.**" So true with math!

The math placement test at each college gives accurate information about the placement in a math course best suited for each student. Repeating previous courses to relearn skills can be a satisfying experience that gets students up to speed and enables them to succeed as they advance from course to course. Enrolling in the next level without relearning the basics can be very frustrating and can produce anxiety about math. My daughter, now an environmental engineer with B.S. and M.S. degrees, reports that *repeating* the first two semesters of calculus formed the best possible basis for her continued courses and work in engineering.

Total honesty with yourself regarding your math skills is essential to good performance and understanding. Bluffing yourself into thinking you remember or know more than you do only wastes time in the long run. The more honestly you can admit what you do know and do not know, the lower your level of anxiety will be.

Reframe your math work. Change from saying "I should skip to the highest level math class" to saying "By taking math courses at my skill level, I will complete my math requirements more quickly than if I jump in over my head."

10. My Negative Math Memories Will Never Go Away

Brain researchers have discovered that memories are not set in stone—they are fluid and changeable. You revise your memories all the time. For you to grow beyond any negativity in math, your new math experiences *must be different* from the old ones.

(*Note:* If you have very vivid and emotionally charged negative memories of math experiences, you may wish to get some professional assistance from a counselor to *separate* the negative experience from the math content. Many community colleges and universities provide this service at no cost to students.)

As you read this book and gain a broader perspective, you will know more about the roles both you and your teacher play. You will also know when bad experiences come from your own avoidance behaviors, lack of information, or acting helpless. You will learn when it is valid to say: "Excuse me. I believe this negativity belongs to someone else," *or* when it is in your best interests to say: "I think I need to act in a different way to achieve success." Chapter 6 will give you suggestions for finding positive encouragement and support for your math work.

To revise negative memories about math, first acknowledge the bad experiences and look at the people involved as well as your behavior. It is likely that any insensitive people involved lacked information or were limited by their own math anxiety. Their learning style might have been different from yours and they could not overcome this difference to communicate with you. As they tried to teach you, they felt responsible for your learning and helpless to communicate at the same time. Feeling both responsible and helpless can cause some people to react with anger.

A teacher, parent, tutor, or fellow student may have said or done some very negative things to you *but* if you now continue to repeat over and over what was said, you are choosing to reinforce the negative feelings. **This is the point where you can now intervene and take responsibility for change.** You can recognize that those people were wrong and replace those negative messages with new messages that are positive for you. As artist/writer Brian Andreas said, "I once had a garden filled with flowers that grew only on dark thoughts, but they need constant attention and one day I decided I had better things to do."

Reframe your math work. Change from saying "My negative math memories will never go away" to saying "Negative math memories can fade and be replaced in time with current positive experiences that I control and choose."

This book will help you create new experiences that highlight your abilities to change and to grow. You might be surprised at what releasing the old tapes in your mind can accomplish!

POSSIBLE SHORT-TERM GOALS TO CHANGE YOUR MATH BELIEFS

1. Write three ways that you approach math differently now than in the past.
2. Copy the reframes for all the misconceptions you marked true in the quiz.
3. Revisit the T/F quiz weekly to review these misconceptions. Remind yourself that they are false.
4. Ask your teacher for other misconceptions about math.

ACT FOR SUCCESS | CHAPTER 1

1. Write three actions for studying math successfully that you have learned from reading the explanations in this chapter. Explain how these actions help math students. (Example: Enroll in the right level of math. This helps math students because their frustration level will be lower in the right level of math, making it easier to learn and to persist.)

2. List any beliefs from the true/false quiz that you marked true. Write down new information that helps you disagree with your old beliefs. Did I convince you to change your mind? If you find a particular "negative math thought" that you still believe to be true, discuss it with instructors and other students.

3. Ask three math teachers or students what misconceptions about math they think unsuccessful math students believe. Write these misconceptions down and why they are not true. Reframe these misconceptions into more useful words about math study.

4. Begin a journal to chart your progress and track your experiences, thoughts, and brainstorms as you read this book and as you work with math. Purchase a fun notebook or sketchpad to use. Daily write down your thoughts, ideas, goals, feelings, doodles, notes, lists—anything that comes to your mind during the day that might involve you and math or you and learning. Remember: **Change is a process. Change takes time.** Date your entries so you can see your progress. Your journal—a powerful tool for change—will help you become more conscious of your inner thoughts. *Experiment with your journal and see what happens.*

> *I use an artist's sketchpad to doodle, take notes, and copy favorite quotations, as well as to write my personal journal. The writing does not have to be perfect. Sometimes as I write in my journal, I think that what I am writing is trivial and a waste of time. I keep writing. Later I often discover that the writing process has revealed something new or I discover that I feel differently after writing and I am released to go on to important work that I want to do. My journaling provides me with direction and focus.*
>
> *In my journal, I experimented with different-colored pens and pencils and discovered that changing colors and writing tools helped me move beyond my previous limitations. In fact, this experimentation helped me discover some unrecognized talent for drawing despite my image of myself as a nonartist. The drawings beginning the chapters in this book are mine.*

Introducing Master Math's Mysteries

The "Master Math's Mysteries" section at the end of each chapter is designed to expose you to math ideas so that you can practice the math study skills you are learning in this book. Each math topic is presented concretely to help you understand the whys behind the techniques. Try the Mysteries, but if you are not ready for the ideas presented, that is O.K. It means that I have not chosen the math that you can learn at your level. Enrolling in a basic math course or working through a book such as *The Only Math Book You'll Ever Need* by Stanley Kogelman and Barbara R. Heller will provide math material at an appropriate level for you.

If you are currently taking a math class, you may have plenty of mathematical ideas to practice and may wish to only read the chapters in this book, practicing the techniques you learn on your math class homework rather than working the "Master Math's Mysteries" section.

Math is not a competition. Take all the time you need with each topic. The solutions to the exercises are in the back of the book.

 MASTER MATH'S MYSTERIES

Increase Your Number Sense with Compatible Numbers

One of the myths about mathematicians is that they do all math problems quickly in their heads. The reason so many people believe this myth is that they see math teachers do arithmetic problems quickly–almost as if they have a calculator in their brain. Well, the truth often is that the math teacher understands certain number relationships that help her form strategies for adding and multiplying more easily. Many students have not been shown these relationships and so what the teacher does seems like magic. Here are some of those strategies for adding and multiplying using relationships between number pairs called "compatibles." (A children's book, *The Grapes of Math* by Greg Tang, shows other strategies for adding. Read it to your young friends and help them with their math. Your teacher can suggest other resources.)

(*continued*)

Adding Compatibles

Certain numbers are easier than others to add together. We call pairs of numbers that add together easily "compatible pairs." For instance, adding 27 + 86 requires carrying, and most people would have to write the problem down to find the answer. Another pair, 25 + 75 = 100, is easier to add. No carrying is required if you think in terms of money. One quarter and 3 more quarters gives you 4 quarters, or a dollar. The numbers 25 and 75 are called a compatible pair.

Circle the compatible pairs below. The first set has compatible pairs adding to 100, and the second set has compatible pairs adding to 1,000. An example has been circled for you in each set.

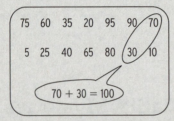

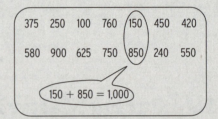

Let's make this strategy more useful. To add two numbers that are not compatible, you can find a compatible pair first, then tack on the remainders. Look at these examples:

$$407 + 600$$
$$= 7 + 400 + 600$$
$$= 7 + (400 + 600)$$
$$= 7 + 1{,}000$$
$$= 1{,}007$$

$$154 + 852$$
$$= 4 + 150 + 850 + 2$$
$$= 4 + (150 + 850) + 2$$
$$= 4 + 2 + 1{,}000$$
$$= 1{,}006$$

Try these by splitting each number so that you can form a compatible pair:

1. 452 + 550   **2.** 1,256 + 753   **3.** 375 + 525   **4.** 903 + 110

5. 406 + 120   **6.** 350 + 650   **7.** 597 + 1,410   **8.** 175 + 330

Multiplying Compatibles

We know that some multiplication is simpler than other problems. For instance, multiplying by 2 or 10 is easier than multiplying by 7 or 9. For some of these nice problems we can find strategies to help in multiplying. For instance, $25 \cdot 2 = 50$ (thinking in terms of money, 2 quarters is worth 50 cents). Also $17 \cdot 10 = 170$ (in terms of money, 17 dimes is worth 170 cents). "Master Math's Mysteries" in Chapter 2 will give you some suggestions for making multiplying by 7 or 9 easier.

Try these: **9.** $75 \cdot 2$ **10.** $12 \cdot 10$ **11.** $15 \cdot 2$ **12.** $10 \cdot 42$

Hints: Think quarters.   Think dimes.   Think nickels.

Multiplying and Adding Compatibles

Now let us combine adding and multiplying compatibles. For example, $26 \cdot 2$ means 26 sets of 2. Because 26 is one more than 25, $26 \cdot 2$ is the same as $(25 + 1) \cdot 2$, and we can think of

$26 \cdot 2$ as 25 sets of 2 plus one more set of 2. So, the problem becomes as follows using compatible numbers:

$26 \cdot 2$
$= (25 + 1) \cdot 2$
$= 25 \cdot 2 + 1 \cdot 2$
$= 50 + 2$
$= 52$
Note that $25 \cdot 2 = 50$, then add on one more 2.
Therefore, $26 \cdot 2 = 52$

Try these to practice using addition with multiplication:

13. $53 \cdot 2$ **14.** $4 \cdot 26$ **15.** $11 \cdot 5$ **16.** $12 \cdot 5$

More Multiplication

Here is another multiplying strategy that mathematicians use. Let's say that you are multiplying by 5. We know multiplying by 10 is easier. We also know that $5 \cdot 2$ is 10. So if we multiply 5 by an even number, we can make the problem easier. For example, consider: $5 \cdot 14$. In terms of money, one way that we can think of this problem $5 \cdot 14$ is as 14 nickels. But every 2 nickels is worth a dime. So instead of 14 nickels we have 7 dimes which are worth 70 cents. In mathematical terms, $5 \cdot 14$ is $5 \cdot (2 \cdot 7)$, but we can think of this as $(5 \cdot 2) \cdot 7 = 10 \cdot 7 = 70$.

Try these: **17.** $12 \cdot 5$ **18.** $26 \cdot 5$ **19.** $5 \cdot 42$ **20.** $84 \cdot 5$

Student Success Stories

You will meet former Santa Ana College math students featured between chapters and quoted within chapters throughout the book. If you had asked these students when they first came to Santa Ana College if they had ever thought they would eventually conquer and tutor math, they would all have replied, "No way!" When they started Santa Ana College, they took their math classes because they had to, not because they wanted to.

Four of these students had to repeat their math courses. Yet they not only achieved their math goals, but they also became instrumental in the math classrooms and Math Study Center. They became highly valued peer math tutors because of their enthusiasm and their acquired ability to explain and answer questions about mathematics. Students turned to them for feedback and were rewarded for their efforts. Touched by the magic of math, these students, in turn, touched others. In this book, they share with you their experiences and insights in learning math.

Student Success Story

Alex Solano

Disliking math in the beginning, **Alex Solano** now likes how math can be applied to the real world. He said, "What is amazing, and what I never thought about, is how much you can do with math. The higher up I have gone, the more opportunities I see with math. I really like how you can measure certain angles if you are given certain information." Planning to become a kinesiologist, Alex spent the day in shock when his math teacher asked him to be her teaching assistant. During his first year at Santa Ana College, Alex disliked math even though he went to class daily and did all of the homework. Intermediate algebra overwhelmed him, forcing him to repeat it. During the second time through, Alex discovered the patterns and his own love for math. He earned his first A's in math in intermediate algebra, trigonometry, and precalculus, and he credits the strong algebra background he gained from repeating intermediate algebra with helping him understand and pass higher-level courses. The teachers that Alex worked with trusted and valued his assistance with math students.

## CHAPTER 2

Flow with Math

"The quality of life is much improved if we learn to love what we have to do."

MIHALY CSIKSZENTMIHALYI

The Climb

Imagine watching a rock climber. Notice how carefully she examines the route ahead to make choices for the next move. She has four contacts with the rock—with her two hands and her two feet.

A basic rule for rock climbing is to move only one contact at a time. With one hand or one foot, she experiments with the next move that will take her up and keep her secure. **The important issues for her are controlling the situation and continuing to move along a route with positive options.**

Be like the rock climber. Take control, keep your solid contacts with the ground, and move in directions with positive options. Every once in a while, glory in looking back to see how far you have come.

Then this will be a safe journey for you because you will choose the rocks that you are willing to climb and you will take charge of each moment. Your journey will progress one move at a time. Your focus will be on the "here and now" as you choose from the currently available possibilities.

Focusing on the *present moment*—right now—is the best way to take charge of and to make changes in your math studies. Often students focus on future math challenges or past math difficulties. This chapter will teach you how to focus on the *present math moment* and how to create more satisfaction and success with your work.

Flow

You can learn to get "in the zone" with math. Those are the moments that are intensely focused and timeless. Those moments include effortless concentration, detailed interest, and a certain satisfaction. Psychologist Mihaly Csikszentmihalyi (me-high chick-sent'-me-high-ee) calls those moments "being in flow." Being in flow with math is another name for concentrated and clear focus on math.

Athletes refer to this type of experience as being "in the zone." **Musicians** experience "flow" as they perform exquisite pieces of music; **artists**, as they create in their studios. Most of us experience "flow" when we do something we love.

Math student Isabella Vescey experiences "flow" and says she likes math when "math is working for me—that's when I'm comfortable doing it. It's flowing along nicely. I forget about everything else in my life and I'm just right there in the math. I like the feel. My mind doesn't wander. I'm able to pay attention. In statistics, I noticed that I was paying attention rather than wandering off. That kind of amazed me."

Professor Csikszentmihalyi studied thousands of people and found that "almost any activity can produce flow provided the right elements are present." Paradoxically, he found that even though adults and teenagers complain about work or study, they experience flow more often during work or study than during passive leisure activities such as watching TV.

Would you like your math work to be: Effortlessly concentrated? Intensely focused? Full of detailed interest? Satisfying? Timeless? Read on to learn how you can get in flow or the zone with math using three elements.

Elements That Create Flow

1. Match your math skills and the challenges of the math work continuously.

2. Set clear intentions or short-term goals for all math activity.

3. Get relevant and immediate feedback on how you are doing the problems and understanding the concepts.

1. Match Your Skills and the Challenges Continuously

To create flow for yourself, be certain that **your math skills match the demands of your math work at all times.** This is a continuous and dynamic balancing act. High skill and high challenge together are the most satisfying. If your skills remain the same and the challenges increase, you become overwhelmed and anxious until you learn new skills to obtain flow again. If your skills increase and the challenges do not, the balance tilts toward boredom. This means you need to increase the challenge level to get back into flow. Flow is that place between boredom and anxiety where your focus is effortless and concentrated because your skills and challenges match each other. Only you can know and control that match for yourself.

Math courses become difficult if you don't recognize how they are different from other subjects and adjust to that difference. Each math concept depends on its prerequisite concepts. This is different from other subjects such as English or psychology or history where you can often miss ideas or even whole chapters and still understand what is going on. To do well with math, find the right level by taking a placement test and consulting a teacher or counselor and then do whatever is needed to stay at the right level. This level is exactly where your skills meet the challenges. Math courses are skills courses in which performance at a specific skill level is expected throughout the course. College-level math courses are increasingly demanding. Make certain that your math abilities balance these challenges. This means that not only must you make certain that you place yourself at the right level of math initially, but also that you must keep yourself at the right level day to day, almost moment to moment.

If you have not studied the prerequisite course recently or did poorly in the prerequisite course, your current skills will not meet the demands of the work. It is no secret to math teachers that students who passed the prerequisite course with a C grade seldom pass their current math class *unless* they make a major effort to shore up their weaknesses.

Once committed to a level and course of math, it is important to face daily challenges by raising your skill level. This means setting aside enough time to study and ask questions as well as developing a program to practice and review the concepts to keep them current. Students who are frequently late for or absent from class will soon find that their math skills do not meet the demands of the math work because they have missed part of the development of new skills. Even missing 10 minutes at the beginning of a class period can hinder understanding.

Unfortunately, many students begin at the right level but they coast at first instead of doing the easier problems as warm-ups. Then, when they face more challenging work, they are not in shape and become overwhelmed. On exams, some speed may be required to finish on time. Those students who don't repeatedly practice the problems—both easier and harder—may not work quickly enough to complete the exam.

Enrolling in a math class beyond your skill level or not doing the necessary daily work to raise and keep your skills is guaranteed to cause anxiety. **The key to speeding up in math is to slow down and understand the concepts as you move forward. And then, practice.**

Overall, you will save time by taking all of the math courses needed for your major in sequence, one semester after another, because you will maintain your skill level and not have to repeat.

2. Set Clear Goals

Set clear short-term and long-term goals and intentions for all math activities. Professor Csikszentmihalyi says, "[You] can set goals for even the most despised task."

Goals such as earning high grades or completing degree requirements are essential, but they are far from your day-to-day experiences in your math class. If they are your *only* two goals, they take away from the concentration necessary for working math. Each time you don't understand a concept or become confused, you will feel at odds with your future long-term goals. **Set more *immediate or short-term goals* along with long-range goals to bring your attention and intention to the math process going on right now.**

The short-term goals suggested in this section are achievable and measurable daily. They support the broader goals of receiving a high grade and completing the course. They shift your attention and intention to *accomplishment now.* After all, how could you eat an elephant? One bite at a time.

To bring satisfaction with math into the present, set short-term, achievable goals frequently. **At any time you become frustrated or overwhelmed, pull yourself back to the present moment to discover what small, possible goals you could set and achieve within the next 20 minutes that will blend into your larger goals for your education.** These goals could be as simple as:

1. Copy the problem onto a clean sheet of paper.
2. Rework three problems you have completed previously that are similar.
3. Locate the corresponding section in the textbook and reread it from the beginning.
4. Make an appointment to get any needed assistance.
5. Take a three-minute break out of doors.

I have suggested many short-term, achievable goals in the following box as possibilities for you. Make a copy of this list and consult it to set short-term goals if you get overwhelmed or lack focus. Throughout this book I will insert other lists of possible immediate, short-term goals that fit the contents being described. Experiment with these suggestions and with developing your own. I recommend that you daily set three or more short-term math goals that are achievable for you.

SHORT-TERM GOALS

These goals get you in flow with math. Check those that you might consider.

❏ Work three review problems each day to boost confidence.

❏ Summarize what you learned in class today.

❏ Write five questions on what you don't understand in this chapter.

❏ Recognize how much more math you know now than you did two weeks ago.

❏ Review your class notes for last week's material.

❏ Quiz yourself on last week's work using two problems from each section of the textbook.

❏ Write down two problems each day that could be on the next math exam.

❏ Attend class daily on time.

❏ In class, mark in your notes what the teacher considers important.

❏ Copy everything the teacher writes on the board into your notes.

❏ Practice patience with your understanding of new processes. Breathe deeply to relax.

❏ Mark where you don't understand your notes and textbook and ask questions.

❏ Complete 90% of the assigned homework with understanding.

❏ Make an appointment to get help from a teacher, tutor, or counselor.

❏ Speak personally to the teacher to establish rapport and increase your comfort level in class.

❏ Introduce yourself to three classmates to develop a math support system.

❏ Start or attend a study group.

❏ Bring problems to class with questions. If they don't get answered, find another way to get answers.

❏ Turn in completed homework on time.

❏ Begin math study and homework problems within three hours after class.

❏ Notice your automatic negative math beliefs and read how to reframe them in Chapter 10.

❏ Study in the math tutoring center.

❏ Cheer yourself and your fellow students as you learn new ideas. Smile and laugh in math class.

3. Get Feedback

The last element of creating a flow experience with math is to **get relevant and immediate feedback on how you are doing problems and understanding concepts.** Because of great individual differences, you will, no doubt, see math problems differently from how others, including teachers, see them. For that reason, to understand math *completely,* continuous feedback is essential.

Feedback is clear and timely input on whether you understand correctly or not. Feedback shapes your growing understanding as you learn. It is information on how well you understand the ideas, concepts, and problems in your math work. Without feedback you will not know whether your understanding is correct or not. Possible activities to provide feedback are suggested on the following page.

Often we hear the comments of others about our work as criticism. Reframe their comments as feedback and welcome them. I learned to appreciate the red marks on my written work when I realized that those remarks could make me a better writer. Those red marks were great gifts to me, and I became grateful for anyone willing to take the time to read my work and give me feedback.

The most powerful way to provide feedback for yourself is to teach someone else. As you teach and explain, you will be considering the problem differently as you seek the clearest way to help the other person understand. You will also be verbalizing the problem rather than just thinking about it. Different and more parts of your brain will be involved. Brain researchers know that this increases understanding. Many teachers confess that they learn their subject really well only as they teach it.

Lots of feedback is necessary to logically order your learning in your mind. Only you, yourself, can be certain that you get enough. Even teachers need feedback. An outstanding math teacher once told me how she got feedback herself. When she was teaching her class about right angles (those angles with square corners), she always drew them with the horizontal leg extending to the right. One day she drew an example with the horizontal leg extending to the left and the students said that was a "left angle." That student feedback caused the math teacher to change her method of teaching, and she wondered how many of her previous students still didn't understand what a right angle was.

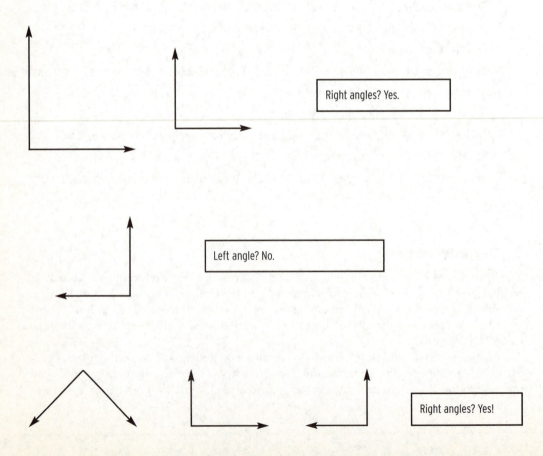

FEEDBACK ACTIVITIES

These actions provide feedback about whether or not you understand. Check those you might consider.

- ☐ Teach someone else how to do your homework problems.
- ☐ Work the examples from class over until you can do them without consulting your notes.
- ☐ Solve a problem several ways.
- ☐ Ask yourself if the answer makes sense.
- ☐ Work through the original problem using your answer.
- ☐ Check your answers against the answers in the back of the textbook.
- ☐ Ask questions about your work.
- ☐ Summarize in writing the procedures of the solution and draw a picture of the problem.
- ☐ Tutor other students.
- ☐ Share your work with the class when students are asked to work at the board.
- ☐ Show your work to someone who knows how to work the problem. Welcome their corrections as feedback, not criticism.
- ☐ Talk over a problem with a tutor.
- ☐ Visit the teacher during office hours and ask questions.
- ☐ Consult your study group and discuss problems with classmates.
- ☐ Answer questions in class asked by the teacher and other students—if only to yourself.
- ☐ Copy the examples the teacher gives in class; then work them on your own without looking at your notes.
- ☐ Work through examples in the textbook on paper without looking at the book and then compare.
- ☐ Guess the next step the teacher will do during class before she does it.
- ☐ Solve a problem with a group at the board.

Getting in Flow Creates a Positive Math Cycle

To bring focused attention to each math moment and set off a positive cycle, use the elements of flow (Csikszentmihalyi, 1997).

1. Make sure your math skills and the challenges of your math work match daily.

2. Set immediate short-term goals for your math activities.

3. Get all the feedback you need to understand the problems and concepts.

Consciously working for flow can create a positive math cycle, as shown in the nearby illustration. Focused attention causes increased success, which causes more positive feelings, which creates sharper interest, which causes focused attention. And around the cycle you go!

ACT FOR SUCCESS | CHAPTER 2

1. Describe and define "flow." Write down experiences in your life when you experienced flow.

2. Honestly, have you placed yourself in the right level of math so that your skills match the course demands? Your college's placement test, a math teacher, or a counselor can help. If you are at the correct level, write the actions you take to continue the match.

3. Write four immediate math goals you can achieve this week. At the end of the week, evaluate whether or not you achieved your goals.

4. Think about how you get feedback or input on how you're doing in math. Write six ways that you can increase the feedback you're getting.

MASTER MATH'S MYSTERIES

Counting and Multiplication Tables

You do math when you drive, spend money, and *count*. Depending on your life experience, your math skills differ from others'. Practice a few of your number and multiplication skills now. Check your answers with the solutions in the Appendix.

1. Write the counting numbers from one to nine.

_____ , _____ , _____ , _____ , _____ , _____ , _____ , _____ , _____

The nine symbols that you wrote above and zero are the basis of our number system. There are likely 10 of these symbols because we have 10 fingers and 10 toes. (A bit of trivia: If we had eight fingers and eight toes, our number system would probably have only eight symbols, leaving out the 8 and the 9. Counting with eight symbols could look like this: 0, 1, 2, 3, 4, 5, 6, 7, 10, 11, 12, 13, 14, 15, 16, 17, 20, 21, . . . Weird, right?)

2. Sometimes we count by twos like this: 2, 4, 6, 8, . . . Write the next six numbers of this sequence after 8:

  2 , 4 , 6 , 8 , _____ , _____ , _____ , _____ , _____ , _____

3. Sometimes we count by threes. Count by threes from 3 to 30. Complete the sequence below:

___3___, _____, _____, _____, _____, ___18___, _____, _____, _____, ___30___

The sequence that you just wrote could help you remember multiplication by three. Here's how: Count the sequence aloud using your fingers. To multiply 6 times 3, say 3, 6, 9, 12, 15, 18, keeping track on your fingers until you come to the sixth finger. The answer to 6 times 3 is 18. (If you are hesitant to use your fingers, reread the eighth myth in Chapter 1.)

4. Count by fours from 4 to 40.

___4___, _____, _____, _____, _____, _____, ___28___, _____, _____, ___40___

Do you have trouble multiplying by four? Practice saying the sequence above aloud using your fingers. Raise another finger with each number to cement these numbers in your mind. Multiply 7 times 4 counting by 4s on your fingers. Notice that 28, the answer, pops up with your seventh finger.

5. Count by fives from 5 to 50. Write the numbers here:

___5___, _____, _____, _____, _____, _____, ___35___, _____, _____, _____

To help you learn multiplication facts, practice saying the sequences above aloud or writing them counting by 2s, 3s, 4s, and 5s. If you say them enough, you will memorize them. When you have learned these sequences, you can use your fingers as suggested above for multiplying by the numbers 2, 3, 4, and 5. For example, to multiply 8 times 4, say the sequence of 4s aloud (4, 8, 12, 16, 20, 24, 28, 32) raising one finger for each number. When you reach your eighth finger, you know that 8 · 4 is 32. If you need to multiply 4 times 8, remember that this is the same as 8 times 4. Another example: to recall 7 · 5, count aloud 5, 10, 15, 20, 25, 30, 35 until you come to the seventh finger. The answer is 35.

USE YOUR CALCULATOR IN A NEW WAY

Another way to practice multiplication facts is to use a calculator. Punch 7 · 4 into the calculator *but* guess the answer before punching "equals." Then punch "equals" and *just notice* whether your guess was too big or too small. Practice this way first working with your 3s, then 4s, then 5s. Mix up the problems. For example, try 3 · 6, then 4 · 7, then 2 · 6, and so on. By the time you get to problems involving only 6, 7, 8, or 9, you will have only a few new facts to learn. Use the calculator method for practice during the little bits of spare time that you have during the day.

(*continued*)

Once you know the sequences for 1 through 5, only 10 more multiplication facts are left in order to know the facts from 1 through 9. The missing facts are these:

6s	$6 \cdot 6 = 36$	$6 \cdot 7 = 42$	$6 \cdot 8 = 48$	$6 \cdot 9 = 54$
7s		$7 \cdot 7 = 49$	$7 \cdot 8 = 56$	$7 \cdot 9 = 63$
8s			$8 \cdot 8 = 64$	$8 \cdot 9 = 72$
9s				$9 \cdot 9 = 81$

To learn these 10 facts, I recommend that you use your calculator in the way suggested above. Punch in a problem such as $7 \cdot 7$ but, before you hit the = sign, guess the answer out loud, then hit the = sign. Notice whether your guess is too large or too small. (Expect to be wrong in the beginning. You are only guessing!) With continued practice using the calculator in this way, you will soon know these facts even if you forget, lose, or break your calculator.

The illustrations below show a fun finger technique for multiplying by 9.

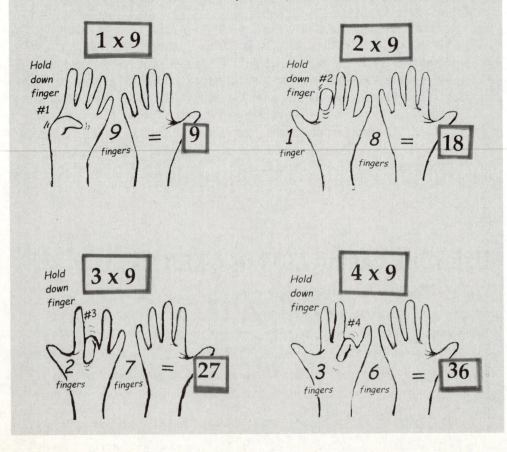

POSSIBLE SHORT-TERM GOALS TO GET INTO FLOW PRACTICING MULTIPLICATION

1. Use the calculator to practice multiplication tables for 5 minutes each day.
2. Write out the sequences for counting with 2s, 3s, 4s, and 5s twice each day.
3. Practice the 10 facts found in the previous table for 6s, 7s, 8s, and 9s using your calculator.
4. Take a daily timed multiplication quiz for two weeks. Use graph paper to make a bar graph of your scores to see how they increase.

MULTIPLICATION QUIZ FOR PRACTICE

Time yourself for three minutes and answer in any order you wish. Answers are in the Appendix. Record your daily scores. Do not expect to finish the first day. Notice that every day you will gain speed and accuracy. Ask a math teacher where you can obtain other quizzes, or make other quizzes yourself by mixing the order of these problems.

$2 \cdot 3 =$ ___	$5 \cdot 4 =$ ___	$5 \cdot 1 =$ ___	$4 \cdot 3 =$ ___	$6 \cdot 6 =$ ___	$6 \cdot 5 =$ ___	$5 \cdot 7 =$ ___
$9 \cdot 3 =$ ___	$4 \cdot 8 =$ ___	$6 \cdot 9 =$ ___	$8 \cdot 8 =$ ___	$7 \cdot 3 =$ ___	$4 \cdot 4 =$ ___	$8 \cdot 3 =$ ___
$6 \cdot 4 =$ ___	$5 \cdot 9 =$ ___	$2 \cdot 9 =$ ___	$7 \cdot 4 =$ ___	$6 \cdot 0 =$ ___	$1 \cdot 8 =$ ___	$4 \cdot 9 =$ ___
$1 \cdot 7 =$ ___	$9 \cdot 8 =$ ___	$3 \cdot 9 =$ ___	$7 \cdot 7 =$ ___	$2 \cdot 7 =$ ___	$8 \cdot 2 =$ ___	$6 \cdot 1 =$ ___
$9 \cdot 9 =$ ___	$4 \cdot 2 =$ ___	$2 \cdot 5 =$ ___	$1 \cdot 4 =$ ___	$5 \cdot 5 =$ ___	$2 \cdot 6 =$ ___	$0 \cdot 8 =$ ___
$5 \cdot 8 =$ ___	$7 \cdot 8 =$ ___	$6 \cdot 7 =$ ___	$3 \cdot 0 =$ ___	$1 \cdot 2 =$ ___	$8 \cdot 9 =$ ___	$3 \cdot 7 =$ ___

Student Success Story

Isabella Vescey

Isabella Vescey likes the feeling she gets when math flows for her—when she's comfortable doing it. "There are times," she says, "when I do my homework and everything else is gone." She confesses that sometimes, while her family is watching television, she sits in the kitchen playing with the numbers and buttons on her calculator. A returning student and a 40-something mom, Isabella used to feel tearful and turn off during beginning algebra classes. In fact, she took the course four times before she passed. In her statistics class, Isabella shocked herself when her hand shot up after her teacher asked who would be taking more statistics beyond the current course. Isabella recognizes in retrospect that she always could do math, but she did not realize that when she was younger. She now attributes her success to not being afraid to take risks on paper and to try new things. As a valued and candid math teaching assistant, Isabella gained work experience that helped her obtain her bachelor's degree and her vocational education teaching credential.

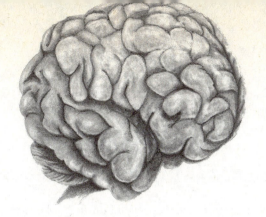

Learn How You Learn

"A successful life is the unique invention of the person who lives it."

ANONYMOUS

You may have answered the question "Who can do math?" many times and left yourself out. I believe, "*Everyone* can do math" and I know that I am not alone. Statistician/Professor David Drew says,

> Research is revealing that mastery of mathematics may be the single factor most related to an individual's success in college and beyond. Furthermore, virtually everyone can learn advanced mathematical concepts, even those who start late. The negative attitudes about mathematics achievement are based on incorrect assumptions about who can learn this subject. (Drew, 1996, p. 9)

Although we do have different ways that we learn best, as a Chinese proverb says, "There are many paths to the top of the mountain, but the view is always the same."

This chapter includes two models to assist you in knowing more about your rich and complicated abilities to learn. These models—Multiple Intelligences and Learning Modes—will help you to identify your personal strengths and match them to your math study strategies.

As you read each section, consider yourself. See what fits you comfortably at this time. When you have thoughtfully completed this chapter, you will have an awareness of many new math strategies that might work for you. Using these models can also help you set effective short-term goals to keep your math skills matched with your math challenges so that you stay in "flow" with math.

A note of caution: Be careful of labeling yourself. Labels are static. You are not. If a label helps you become a "bigger" person with more alternatives, use it. When it does not benefit you anymore, shed it as a butterfly sheds a cocoon.

Model #1—Multiple Intelligences

An innovative and widely referenced model of eight human intelligences developed by Harvard Professor Howard Gardner (1993) is known as Multiple Intelligences. Gardner's work changed the question "How intelligent are you?" to "How are you intelligent?" You are not smart in just one or two ways—but in many ways. Even more, intelligence is not fixed—it can change. This section will show you how to use your multiple, changing intelligences to improve your math.

To develop this model, Gardner looked at the ability to solve problems and produce meaningful products within a person's own culture. He discovered at least eight kinds of intelligence as he studied diverse cultures throughout the world to define intelligence in a manner useful for all types of human beings, whether they are artists, bush pilots, doctors, mothers, athletes, musicians, teachers, sailors, tribal chiefs, or engineers.

The diagram on the next page shows those eight intelligences of Gardner's model. If you are not familiar with these intelligences and have not previously assessed your strengths, read about each in the Appendix of this book and use the simple assessment found there to identify your strongest intelligences at this point in your life. Use that information to discover how to use your strengths with math.

How to Use Your Multiple Intelligences in Math

For those whose strongest intelligence is not Logical-Mathematical, read on for suggestions. Remember: Intelligence can change. Using any suggested strategy listed here that interests you will strengthen your math skills.

■ If you are strong in *Bodily-Kinesthetic Intelligence,* you will find movement useful in your math studies. Memory is triggered by location, so moving your body to different places as you learn new concepts helps you remember them. The gift of this intelligence will help you to lay out or act out problems physically using your muscles and movement. Even moving randomly as you think over, discuss, and work on math problems helps you learn better. Study your note cards of concepts, vocabulary, and problems as you move. Writing on a large chalkboard or whiteboard facilitates that movement. Walking a large figure-eight shape (Sunbeck, 1996) (used by learning disability specialists) as you speak math aloud will cement learning because this movement changes your state of mind, thus involving more functions in your brain.

FACTS ABOUT MULTIPLE INTELLIGENCES (GARDNER)

- Intelligence can and does change!
- All of your intelligences work together.
- Rarely does anyone exhibit only one of the intelligences.
- There is now so much to learn that no one can learn it all, but there are necessary effective living skills that everyone should know, such as math.

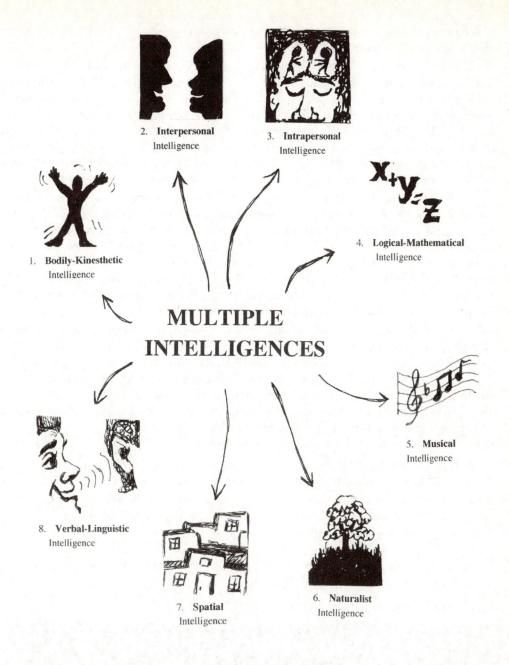

2. **Interpersonal**
Intelligence

3. **Intrapersonal**
Intelligence

1. **Bodily-Kinesthetic**
Intelligence

4. **Logical-Mathematical**
Intelligence

MULTIPLE
INTELLIGENCES

5. **Musical**
Intelligence

8. **Verbal-Linguistic**
Intelligence

7. **Spatial**
Intelligence

6. **Naturalist**
Intelligence

■ If your *Interpersonal Intelligence* is strong, you would do well organizing study groups and facilitating others working together to discuss math problems. Taking a leadership role, you will feel comfortable and find that you sponge up math skills in your interactions with others. With a group you enjoy, facilitate an open forum where questions are welcomed. You may wish to volunteer to tutor students with fewer math skills. As you do this, notice how your understanding increases as you explain ideas to others.

■ Strong *Intrapersonal Intelligence* means you may enjoy contemplating math on your own at least part of the time. You might enjoy biographies of those famous mathematicians who were solitary and contemplative. Systems of thought and systems of social organization can be described in mathematical terms. You may even discover a way that human actions can be described by a mathematical system. Explore chaos theory—a branch of math that brings order to systems previously thought to be unordered.

■ If you are strong in *Musical Intelligence,* you may find the mathematical descriptions of what happens in music fascinating and revealing. Recognize how symbolic, mathematical, and logical music is. By learning music you have already internalized a whole mathematical system that you use in your mind intuitively to create beautiful mathematical/musical structures. Rhythm in music is an auditory manifestation of fractions. You intuitively and creatively put fractional parts together over and over to make whole measures. As you read or perform music, you create fractions of sound in time. It might be helpful for you to find another musician to tutor or mentor you in math or to work with you as a study partner.

■ Having strong *Naturalist Intelligence* means you notice patterns and relationships. Turn this point of view on math by noticing similarities, differences, and categories. Beginning with the whole and working down into the parts may benefit you. It will be extremely important for you to have a working knowledge of the "big picture." You may decide to invent new names meaningful to you for mathematical procedures or concepts. The biological or geological systems with which you are familiar are structured similarly to mathematical systems. For example, the numbers on the real number line are a system with subcategories, the way plant and animal life are systems with subcategories.

■ If your *Spatial Intelligence* is strong, you learn well by making models of problems— models that you can manipulate and move in order to understand symbolic meaning. This intelligence is basic and essential to the study of math because many concepts relate to three-dimensional figures. Use clay or plastic straws to create geometric shapes used in math problems. Use model cars or airplanes to simulate the action in motion problems involving distances, rates, and times. Visit educational supply stores to see the many varied models and manipulatives available in the math section. Sharing concrete models that you find useful may enhance the mathematical learning of others.

■ If you are strong in *Verbal-Linguistic Intelligence,* you may be more comfortable reading math books that explain with more words and fewer symbols. Write math symbols, formulas, and procedures in words so that you can make sense of them and see and appreciate their beauty. You will be happy to know that the use of words is essential to higher math, in which few symbols are used and ideas are often expressed solely in words.

What short-term goals will improve and bring flow to your math learning? The following list contains examples for different intelligences.

FIND FLOW WITH SHORT-TERM GOALS USING MULTIPLE INTELLIGENCES

1. Incorporate more motion into your studies. (Bodily-Kinesthetic Intelligence)
2. Read about how to form a study group in Chapter 6. Then form or join one. (Interpersonal Intelligence)
3. Notice which of your intelligences you use as you study math. (Intrapersonal Intelligence)
4. Incorporate musical memory techniques into your math study. (Musical Intelligence)
5. Ask your instructor about the real number structures of math. (Naturalist Intelligence)
6. Make and draw models of all the math problems that you can. (Spatial Intelligence)
7. Write difficult math problems out in words to make them easier and memorable. (Verbal-Linguistic Intelligence)

Model #2—Learning Modes

Visual, Auditory, and Kinesthetic

We access and process information from the environment through our eyes, our ears, and our skin and muscles. The Learning Modes model focuses on visual, auditory, and kinesthetic learning and is fairly well known and simplistic. If you are unfamiliar with your own Learning Modes, you may wish to turn to the Appendix to take the Learning Modes assessment quiz and read the explanations of each mode before you read this section.

Often one of the three modes—Visual, Auditory, or Kinesthetic—is primary and is used more than the others. The truth is that you use all three modes to make sense of the world around you. When math is presented in one of your weaker modes, step up your efforts and translate the material to a form you will understand. For example, visual learners need to work harder and to translate to visual information when material is presented verbally.

Use your strongest mode(s) to compensate for your other mode(s). Do not give away control over how material in the classroom is presented to you. Practice adapting classroom material to your own mode when it is presented to you in a different mode. Experiment with your weaker modes to improve them. The following section lists many short-term goals you might adopt to involve your learning modes more actively into your learning.

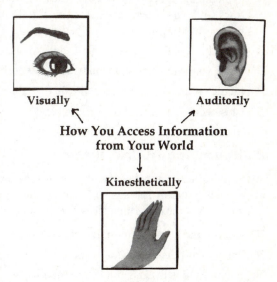

Visually   Auditorily

**How You Access Information
from Your World**

Kinesthetically

As you read the following sections, mark with a check the activities that you habitually use. Mark other activities that you find promising with a question mark. **The more activities you involve in your learning from** *all three* **of the modes, the more easily you will learn.**

How to Improve Visual/Auditory/Kinesthetic Input into Math

1. Improving Visual Input

Visual learners learn more easily when they see the material. If you are primarily a visual learner, any strategy that allows you to see better, see more, or see connections will assist you to learn more effectively. Often you can adapt material that doesn't fit your style. For example, if your teacher lectures, transcribe the auditory lecture into visual input by taking creative, detailed notes.

Here are actions that assist visual learners to set short-term goals for math work. Mark those of interest, use them, and analyze what works.

❐   Sit in the front of classes or meetings so you can see everything.

❐   Make interesting-looking note cards with formulas, facts, vocabulary, and sample problems.

❐   Sketch the course content. Even the crudest sketch can help you remember ideas.

❐   Make note taking fun by using color and little doodles. Embellish the pages or note cards to look nice.

❒ Develop skill at note taking by practicing changing verbal input into visual input.

❒ Mind Map® the course content. (See Chapter 8.)

❒ List your tasks—even the ones you've completed—to have the satisfaction of visually crossing them out. (My husband laughs when I do this, but it works for me. I see that I've done my jobs.)

❒ Use notes on stickies to help you remember. Use your favorite colors!

❒ Evaluate the appearance of your study environment. Make it *look* conducive to learning. A well-placed poster that you love or a desk turned away from clutter may work wonders in clearing your mind to study better.

❒ Quickly clear the mess from your desk to clear your mind for study.

❒ Write yourself encouraging messages and post them where you can see them.

❒ Picture yourself in situations in which you have succeeded.

❒ Close your eyes when you want to block out unpleasantness.

2. Improving Auditory Input

Auditory learners can best remember what they hear. Strategies that improve or stress hearing work well for auditory learners especially when they combine auditory input with activity. Try the following. Mark the interesting ideas, use them to set short-term study goals, and evaluate them.

❒ Choose the best classroom location for listening.

❒ Tape-record the class session and listen to your tape.

❒ Ask questions in class and listen carefully to the replies.

❒ Read the textbook and class notes *aloud* to yourself as you study.

❒ Record and listen to your textbook or your class notes.

❒ Study with others. Talk about the course material.

❒ Tell others (your pets or whoever will listen) what you are learning in class. Mentally replay your speech during exams.

❒ When you study, choose the auditory input in the background carefully. You are influenced by the sounds around you—especially talking. You may discover that you have favorite background sounds that help you concentrate.

❒ Use headphones so that the auditory input is of your own choosing.

❒ Consider using earplugs during exams to mask distracting noises. Inexpensive earplugs are available at drugstores. Notify your instructor in case he makes announcements.

❒ Teach yourself to read aloud in your mind without making sounds. During exams, you can hear the test questions as well as see them. (If you need to move your lips, warn your instructor so you aren't accused of cheating.)

❒ Speak positively to yourself during your work.

3. Improving Kinesthetic Input

Kinesthetic learners learn more easily when their skin and muscles are involved. Motion or activity involving the subject matter will help you learn more effectively. Here are actions to help you set short-term goals that fit you. Mark the interesting actions, do them, and evaluate their success.

❒ Sit where you can actively participate in classroom events.

❒ Sit where you can move as needed without disturbing others in class.

❐ Draw sketches and diagrams in class of the material being taught.

❐ Take notes creatively using different colors. Turn your notebook around and write up the page from the bottom on occasion.

❐ Ask and answer questions.

❐ Make models of the concepts whenever possible. Visit an educational supply store to see mathematical models.

❐ Become skilled using your fingers and toes when doing math.

❐ Learn to calculate on a Chinese abacus.

❐ Educate your instructor about kinesthetic learners and ask for assistance in developing models of the material with which you can interact physically.

❐ Move around as you study your note cards of math facts, formulas, and problems.

❐ Talk to yourself about the material as you walk.

❐ Walk a figure-eight pattern, swinging your arms as you recite material you want to remember for your coursework. This walk will activate different parts of your brain and integrate concepts more fully.

❐ Mind Map the course content. (See Chapter 8.)

❐ Work on the chalkboard or whiteboard whenever you can.

❐ Purchase a whiteboard for your own studies at home.

❐ Pat yourself on the back—physically!—whenever you do well.

❐ Make sure your pen and other writing tools please you.

❐ Make physical comfort a priority as you study.

Whether you are a visual, auditory, or kinesthetic learner, always take charge of your own learning! Write some short-term goals now that will use your strengths.

Examples of short-term goals that incorporate more than one learning mode are suggested in the box.

VISUAL/AUDITORY/KINESTHETIC SHORT-TERM GOALS TO GET INTO FLOW WITH MATH

1. Practice note taking in a more interesting way by using organizational symbols, colored pens, Mind Mapping, and sketches. (Visual/Kinesthetic)

2. Read your class notes aloud as you recopy them in an attractive and organized format. (Auditory/Visual/Kinesthetic)

3. Choose the best position to sit in for you to see, hear, or move. Arrive early and sit there. (Visual/Auditory/Kinesthetic)

4. Purchase a whiteboard for your use at home when you practice examples that were worked by the instructor in class. Talk to yourself aloud as you work the problems. (Kinesthetic/Visual/Auditory)

5. Make math more fun by incorporating doodles into your notes, drawing happy faces on what you understand, making up silly songs or jokes to remember concepts, or working with others. (Visual/Kinesthetic/Auditory)

ACT FOR SUCCESS | CHAPTER 3

1. *List your eight Multiple Intelligences in order from your strongest to your weakest. What are your top three? Which of my suggestions can you use to improve your math smarts?*

2. *Which learning mode do you use the most? List three activities that you have not used before that might fit with your way of learning.*

3. *List 10 actions from this chapter that could bring flow (suggested in Chapter 2) to your math work. The actions could become short-term goals that might bring flow to your math work.*

MASTER MATH'S MYSTERIES

Order of Operations

Math has an understood set of rules that everyone is expected to follow—much like behavior in society is governed by rules that no one mentions except to young children like not picking your nose or pointing at someone. The rules in math are called the "order of operations." The operations (add, subtract, multiply, and divide) must be completed in the same order by everyone—if people are to get the same answer to arithmetic problems.

For example: $3 + 2 \cdot 4$

Some people say the answer is 20 and others say 11. If you add first, you get $5 \cdot 4 = 20$. If you multiply first, you get $3 + 8 = 11$. Which is the answer that all will agree to be the correct one? Only the order of operations will tell us this.

Order of Operations Steps

Start at the top and work down the list. The more powerful items are toward the top.

When there is more than one example from each level, always work from innermost to outermost and from left to right.

Parentheses or Grouping symbols such as ( ) [ ] { } | |	P Step
Exponents or Roots–2^3 and square roots	E Step
Multiply or Divide–Work left to right.	M Step
Add or Subtract–Work left to right.	A Step

In our example, $3 + 2 \cdot 4$, we have to choose whether to add or multiply first. Starting from the Top, Order of Operations says that multiplying is more powerful than adding, so we multiply first. The correct answer to the problem is 11.

	$3 + 2 \cdot 4$
M Step	$3 + 8$
A Step	11

I use the crazy word PEMA to help me remember the steps in the order of operations. Just remember that multiply and divide are on the same step M and add and subtract are also on the same step A. Here are three examples:

Example 1	Example 2	Example 3
$6 + (7 - 4)^2(5)$	$7(3) - 2^3$	$[8(2) - 4] + 9 - (5 - 4)^3$
P Step $= 6 + (3)^2(5)$	E Step $= 7(3) - 8$	P with M Step $= [16 - 4] + 9 - (5 - 4)^3$
E Step $= 6 + 9(5)$	M Step $= 21 - 8$	P with A Step $= [12] + 9 - (5 - 4)^3$
M Step $= 6 + 45$	A Step $= 13$	P Step $= 12 + 9 - (1)^3$
A Step $= 51$		E Step $= 12 + 9 - 1$
		A Step $= 21 - 1$
		A Step $= 20$

Now you try these. (Answers are in the Appendix.)

1. $[7 + 6(3)] - 14$ **2.** $5(6 - 4)^2 \div 2$ **3.** $6 + (3)5 - 3^2$

4. $12 \div 3(10 - 7) + 9$ **5.** $12 - 3(10 - 7) + 9$

The previous two problems can cause problems even for the most experienced student. In problem 4, many students incorrectly multiply before dividing after they simplify inside the parentheses. However, in problem 5, you correctly multiply the 3 by the simplified answer from the parentheses, then subtract the result from the 12. A common error in problem 5 is to first subtract the 3 from the 12. It feels like the $12 - 3$ is set off together as if they are in parentheses, but look carefully–they are not inside any grouping symbols.

Important Vocabulary

Sum, Difference, Product, and Quotient

Learn the Vocabulary: Sum, Difference, Product, and Quotient.   These four words are commonly used for operations in mathematics.

Sum means to add.

Difference means to subtract.

Product means to multiply.

Quotient means to divide.

The four operations (add, subtract, multiply, and divide) are usually learned in that order. First, you learned to add and then subtract. Later, you learned to multiply and then divide. If you remember the words *sum* (means add), *difference* (means subtract), *product* (means multiply), and *quotient* (means divide) in the same order, you are likely to recall the correct operation. Here are some weird sentences, some visual cues, and a song to help you remember the meanings of these four words, which show up everywhere in math.

(*continued*)

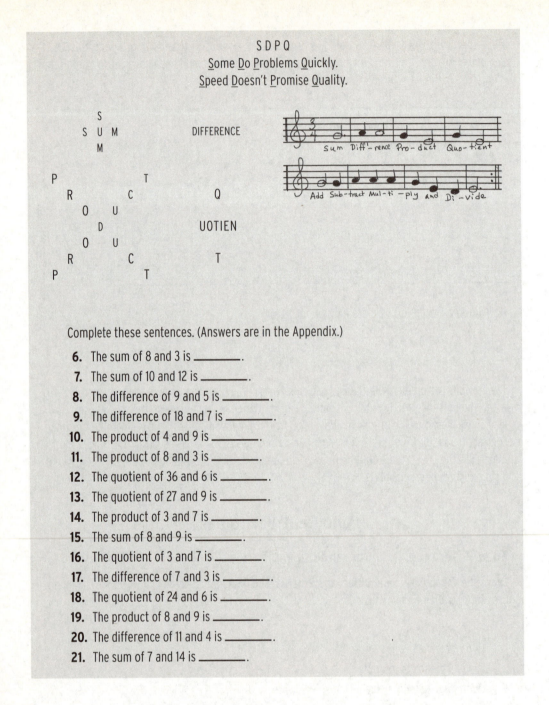

SDPQ
Some Do Problems Quickly.
Speed Doesn't Promise Quality.

Complete these sentences. (Answers are in the Appendix.)

6. The sum of 8 and 3 is _____.
7. The sum of 10 and 12 is _____.
8. The difference of 9 and 5 is _____.
9. The difference of 18 and 7 is _____.
10. The product of 4 and 9 is _____.
11. The product of 8 and 3 is _____.
12. The quotient of 36 and 6 is _____.
13. The quotient of 27 and 9 is _____.
14. The product of 3 and 7 is _____.
15. The sum of 8 and 9 is _____.
16. The quotient of 3 and 7 is _____.
17. The difference of 7 and 3 is _____.
18. The quotient of 24 and 6 is _____.
19. The product of 8 and 9 is _____.
20. The difference of 11 and 4 is _____.
21. The sum of 7 and 14 is _____.

Look Back–Comprehensive Review for Your Practice

To give you more opportunities for practicing and remembering, a cumulative review section follows the "Master Math's Mysteries" section in Chapters 3, 5, 7, 9, and 11. Each "Look Back" contains new problems from the chapter as well as problems from the preceding chapters. The answers are in the Appendix to give you feedback on what you remember.

Try these "order of operation" problems:

    1. $[9 + 6(2)] - 4$ **2.** $5(6 \div 2)^2 - 2$ **3.** $12 + 3(10 - 4) - 9$

Fill in the blank:

    4. The difference of 9 and 4 is _____.

    5. The quotient of 42 and 6 is _____.

Multiply or add using compatible numbers:

    6. $626 + 379$ **7.** $325 + 175$ **8.** $53 \cdot 2$

    9. $4 \cdot 26$ **10.** $11 \cdot 5$

Student Success Story

Enrique De Leon

Student **Enrique De Leon**, when asked what he likes about math, said, "I like going through a process in order to get an answer. The process is actually more fun than the answer. I like the whole struggle of getting somewhere regardless of where it is, and I want to say I love math. It's fun. It's hard. It makes me pull out my hair—when I have some." When Enrique was my prealgebra and beginning algebra student, I frequently gave him my exam answer key after the test to proofread. He was exacting and took pride in detail. Working in groups at the chalkboard in prealgebra class, Enrique took the lead, patiently explaining concepts to other students. Enrique had not thought about teaching math when I recruited him as a Freshman Experience math tutor. By the time he finished his math sequence at Santa Ana College, he had become a popular campus and private math tutor and had a hard time remembering when he didn't want to become a math teacher.

Enrique's advice to you is: "The best thing is to give [math] time. Just don't stop."

Time Yourself Successfully

"If you don't control your attention, much of your psychic energy will be [drawn off] by outside influences."

MIHALY CSIKSZENTMIHALYI

Your time is valuable. With the proven time management activities and practices in this chapter, you can maximize your use of time and minimize the amount of time that you need for math study. But, make no mistake, math study takes time. In this chapter, Santa Ana College math students—Enrique De Leon, Jazmin Hurtado, Sarah Kershaw, Joel Sheldon, Alex Solano, and Isabella Vescey—share with you how they use their valuable time. This chapter will guide your success and help you access flow during those critical times at the beginning of the semester, daily before class, during class, after class, and at the end of the semester. Test time is also critical. Chapter 7 will give you activities to help you before, during, and after testing.

Time Management over All

Begin managing your time for math success with these general strategies.

1. **Make Math a Priority.**   Set math as a top priority so that it gets done. Math will be time intensive, so set aside time for math and say "no" daily to lesser priority activities that might fill it. Fit your other activities around math study. The more time

you focus on math, the better you will get and the more time you will want to spend with it—which makes you more focused. Then you learn the math you need! Math student Isabella Vescey talks about setting priorities and working in her own personal Grand Central Station:

> "Scheduling is a tough one. Lots of times I just have to let a lot of other things go. When I sit down to take the test, I know if I needed to take the time and I didn't take it. One of my best places to do the math is in the kitchen because everything else is there that I need to do—laundry, meal preparation, eating, or answering the phone for my daughter."

2. Make a Schedule and Keep a Calendar. Write out a schedule of the exact times and locations for your math study. Then honor that schedule. A blank schedule is found in the Appendix for your use in planning your time this semester. On this blank schedule, use big categories such as Class, Sleep, Personal Care, Studying, Eating, Commuting, Family, Housekeeping Chores, Work, and so on. Build in flexibility to give math the time that it needs. Also keep a calendar so that you know when you have appointments, exams, papers due, and special events. If it is possible to cut back on your expenses and work fewer hours, do it to give more time to your math course and completing your education. If it is possible to delegate some of your household or family responsibilities during the semester to give you more math time, do so. If your spouse or another family member is willing to make your lunch or babysit or do your laundry so you can study, let them.

Math student Joel Sheldon stresses that scheduling makes things happen even though plans change:

> "I try to make a little schedule like I'm going to try to get this done in this much time. Many times I had ideals of finishing ahead and, with all of my classes, I just didn't. But planning my time made me get it done no matter what."

Math student Sarah Kershaw has found that scheduling large blocks of time for math works best for her:

> "I do better in complete immersion. I find learning little bit by little bit frustrating and confusing. I set aside large blocks of time in which to focus. **It has taken a lot of growing up [for me] to say 'No' to things that interfere in those blocks of time.**"

3. Work in the Best Place. Work math in a location where you are able to concentrate—away from friends and family and distractions. Many students find it easier to study math on campus in either the library or a drop-in study center. Some students find that they can tune out the impersonal noise in the school cafeteria and focus on their math studies. Find the place where you control your attention. Psychologist Mihaly Csikszentmihalyi reports that our psychic energy is limited and is easily drawn off unless we learn to focus. Be in each individual moment. Focus on what you are doing at the time. If you are studying, study. If you are tired, relax and let your studies go for the moment. Make your breaks brief and refreshing.

4. Make Spare Moments Count. Be prepared to make use of available bits of time that appear unexpectedly. If a class is canceled, use that time for math. Take your math note cards or homework problems to the bus stop or laundry or doctor's office. I have read complete math books and worked all the math problems in airports and on airplanes over the course of several trips.

5. Postpone Social Activities. Place your schoolwork ahead of your social life. True friends will understand those priorities and spend time with you when you are available. There will be time for celebrating and seeing friends during school breaks and after you graduate and achieve your educational goals. Enrique underscores the importance and reasons for giving time to math:

> "My advice is to take your time. You won't understand math in five minutes. You won't get it the day before the test. You won't always get it when you're in the classroom. The best thing is to give it time. Just don't stop. I make time. I eat, sleep, and breathe school. So I always have time for school."

Critical Time: The Beginning of the Semester

Get up and running right away. Take charge of success from the beginning to get a head start and lower anxiety. Success builds on success. Here is a winning plan. Use it as a checklist to start your semester.

 SHORT-TERM GOALS FOR FLOW AT THE BEGINNING OF THE SEMESTER

1. **Enroll at your level.** Match your skills to the challenges of your math course by taking the placement exam offered by your school and then enrolling in the advised level. A counselor or math teacher can help you with your best choice for success.

2. **Pick the best time for class.** Select a time for you to take the class when you are the most alert and when you will be able to set aside time for study following the class. Take the class as many times per week as possible so that the course content is spread over several days. Math student Jazmin Hurtado says, "Success depends on my schedule. It's best for me to take a [midday] class every day if I can [in order to] keep [math] in my head." Schedule all of your classes with time between so that you are not pressured toward the end of class and have time to begin homework immediately.

3. **Match your teacher's style.** Choose the teacher who best matches your learning style. Other math students, counselors, and teachers can advise you.

4. **Remember to make the time for math.** Arrange your schedule so that you have time for math. Make math a priority. Limit other activities. Passing your math class the first time will save you money and time. Math is time intensive.

5. **Buy your math book soon.** You will need it immediately in your class and will not want to get behind. Borrow money or a book if you have to.

6. **Assemble your supplies.** You will need writing tools such as sharp pencils, colored pens or pencils, erasers, and highlighters; paper such as graph paper, three-hole notebook paper, and scratch paper; a calculator; a three-ring binder; and a straightedge. Using graph paper for note taking and homework problems organizes those numbers and letters in a line both horizontally and vertically and increases accuracy. Look for my personal favorite graph paper, called "Engineers Computation Pad." It has extra organizing lines and the front side is lighter. Other supplies that I also like to keep at my desk are a stapler, staples, three-hole punch, a tape recorder, blank tapes, and batteries.

(continued)

7. **Organize.**   Set up a math notebook in which you are prepared to file your syllabus, math class schedule, homework assignments, class handouts, class notes, quizzes, past homework, and past exams so that you do not waste time looking for them when you need them. Use divider sheets with tabs labeled with categories so that you can file and find what you need quickly.

8. **Investigate resources.**   Plan your use of resources (classroom, study environment, teacher, tutors, study groups, library, and study centers) ahead of time so you don't waste time looking for them when you are desperate for assistance. Write down specific locations and phone numbers. You might visit the location of each to ask how to use their services. Chapter 6 will give you more suggestions for using these resources.

9. **Increase your comfort on campus.**   Spend time hanging out and getting the lay of the land. Locate the bookstore, administration, faculty offices, cafeteria, restrooms, library, counseling offices, and places to sit and relax. Take a friend to explore campus. Locate your classrooms. Look over the whole campus—including parking.

10. **Set up an educational plan.**   Make an appointment with a counselor to set up an educational plan for your studies and degree program. Arrange your math sequence early in your education because math courses are sequential and the sequence may take several semesters to complete.

11. **Preview your textbook.**   This helps you get a head start by having a sense about vocabulary and where you are headed. Read Isabella's description of her "textbook reconnaissance."

Math student Isabella describes an excellent method for previewing a math textbook:

> "[To preview] I read the summary first and then go back and read the chapter. And now I review the entire math book at the beginning of the semester. They taught us this in study skills. It's called '**textbook reconnaissance**.' You open the book and begin with reading everything on the covers, skim every page at the beginning, read the table of contents, go through the preface and introduction. That gives me a lot of information about what's going to be done during the semester. You learn if there are Web sites, which I used a lot one semester because the Web sites had quizzes. If the book has an introduction to calculators, I would look at that. [D]on't read everything [in the chapters]. Look at headings. Look at drawings. Look at sidebars with applications. Look at everything in bold print. It's a nice weekend activity if no one's around for a couple of hours. You can spend time with your new textbook. [My] algebra book and statistics book had sections on how people use math in their jobs. I remember one on counting moose."

Critical Time: Daily Before Class

Arriving in class prepared with the topics and problems from the last class in mind keeps you in charge and on a successful path because you will understand today's class more quickly and thoroughly. The Santa Ana College math students stress the importance of attendance. Isabella says, "Show up for class every day and show up on time. If you don't go to class, you miss stuff." Joel advises you to stay involved:

> "I go to class every day. You can't just tell yourself that you don't like math class and go every once in a while. You can't not do your homework and give up. You can't not ask your questions. I think that's death and doom right there. I've seen it happen lots of times. I've seen people . . . fade out and disappear."

Check this list for daily prep ahead of class time:

SHORT-TERM MATH GOALS FOR FLOW BEFORE CLASS DAILY

1. **Warm up before class.**   Preview your class notes by rereading them. Work a few examples completed in the previous class as a warm-up activity.

2. **Bring supplies.**   Be prepared with your textbook, notebook, graph paper, writing tools, calculator, and tape recorder in class daily. Without your equipment, you will be distracted and unable to do your job.

3. **Formulate questions.**   Make a list of questions in your textbook, notes, or homework and take it to class. Ideas for questions are found in the "Crucial Skill #3: Asking Questions" section in Chapter 5.

4. **Plan for getting questions answered.**   You could write them on the board before class, tell the teacher that you have questions, make an appointment with a tutor, or discuss them with fellow students before class.

5. **Be there.**   Attend math class every session. Be on time. Alex reports, "I have only missed one class in three years of attending Santa Ana College." Sarah says, "I attend religiously."

6. **Get help.**   Know your teacher, available tutors, and classmates so you have resources when you have questions or need company or motivation. Become acquainted with other students so you can get notes if you are absent, ask questions, or develop a study group. Read Chapter 6 for suggestions how to form a study group.

7. **Maximize your mind.**   Math student Isabella says, "I make sure I eat before class and bring water." Keep your mind alert by caring for your body with sufficient water, food, exercise, and sleep. The mind/body connection is unavoidable. Tired, dehydrated, hungry, or malnourished students are not successful.

8. **Keep on top of it.**   Read the syllabus over often so that you know what the teacher's expectations and schedule are.

Enrique describes his daily math processes:

"Before class, I look ahead at highlighted things in the chapter. I read them. I don't memorize them. I just glance at them, so that when the teacher talks about them in class, they're familiar. Then I say, 'Oh that's what they were talking about.' Everything makes sense. After I preview the material before class and the teacher helps me makes sense of it in class, the homework actually puts it in my head to where I understand it."

Alex explains his classroom preparations:

"I got into a habit of just briefly reading the highlights of the next section that we were going to cover in class. Just looking at the title gave me an idea. I didn't really look at the examples. I . . . looked at the bold letters that mean that's something important. I . . . wanted to see what was going to happen. I didn't want to get caught off guard."

Critical Time: During Class

Focusing on your learning is key. What you do and where you sit during class can maximize your focus and concentration. Over time, math student Jazmin has changed her preference for the best classroom seat. She says:

"I did like sitting in the back in the corner—not in the middle. Now I can't. I have to be in the front. I need to focus. I need to pay attention. In front, I know I won't get distracted by other things that I see."

Isabella is very certain about what she needs during class:

> "I've found that in the past few semesters there are certain people I just can't sit next to. People who are fidgety or have loud, irritating voices throw me off from what I am doing. So I take care of myself. I sit where I'm comfortable with the other students around me and how the lighting is. Sitting in the front works for me, but I also found that sitting in the corner is good for me because there's nobody behind me. I experiment. Some days I feel like being in the corner and some days I feel like being in the front. I change my seat and am in control of where I sit and whom I sit next to. I bring plenty of pencils and my calculator."

Check this list to maximize your time in class:

SHORT-TERM GOALS FOR FLOW DURING CLASS

1. **Be there.** Don't miss a single chapter in the math story of your course. Attend every class!

2. **Sit in the best spot.** Come early and choose your classroom seat carefully. Do not sit by friends or people who distract you. Plan to participate completely. Sit up front to see and hear or sit on the side to move. Sit by successful students focused on math who participate in class and do homework. Their good work and focus will rub off on you.

3. **Get ready.** Before class begins, open your book, notes, list of questions, and homework. Make your writing tools and calculator easily available. When class begins, there is no time to set up. You will lose valuable instruction if you're not fully prepared.

4. **Get questions answered.** Follow the classroom procedure for asking questions, which might be writing the questions you have prepared on the board before class, asking your questions during class or after class, or going to the teacher's office. Bring a prepared list of questions with you.

5. **Stay focused and alert.** Bring your attention back to the classroom whenever it wanders. Don't tune out. If you feel lost, ask a question or take careful notes that you can study and ask about later. Where you sit matters. It is easier to be distracted sitting in the back of the classroom.

6. **Participate actively.** Listen consciously. Ask and answer questions.

7. **Take notes.** Copy teacher comments, problems, and how they are worked. Write down assignments and anything the teacher says about the assignments. Use color or boxes to keep problems and concepts separated. See Chapter 5 for more ideas on note taking.

8. **Imitate the best.** Model your in-class behavior on the best students that you have ever known. You might interview an outstanding student for suggestions for changing your own classroom behaviors. Experimenting will help you find strategies that work.

9. **Stay to the end.** Don't leave the classroom until class is over and you are certain of the assignment for the next class period.

Critical Time: After Class

The time immediately after class presents you with a onetime golden opportunity. The material from class is fresh in your mind. You have just made many, many connections in your brain synthesizing math. Spending time now reinforcing those connections by rereading and reworking your notes and your textbook will keep those brain networks from dissolving in your mind.

As you rewrite your notes now, you will remember what you didn't write down and will be more able to understand and clarify the material. Jazmin tells us, "At the same time I read the chapter, I look at my class notes so I kind of do the whole lecture again in my head while I'm reading it. I put those two things together at the same time. That helps me to understand the chapter more and the notes better."

Check this list for the best use of your time after class:

 SHORT-TERM GOALS FOR FLOW WITH MATH AFTER CLASS

1. **Know the assignment.**   Don't leave the classroom until you have written a copy of the assignment for next time.

2. **Rework your notes.**   Reread your class notes immediately and fill in the missing pieces. Then copy them as soon as possible.

3. **Reread the textbook.**   Right after class you will understand the material you have just heard better than any other time, especially if you previewed the text and have reworked your class notes. Time spent with your textbook now using pencil and paper will make your homework problems easier and more understandable. **Write as you read.** Expect to read your math book with pencil and paper handy to work examples. Write in the book. See Chapter 5 for more ideas on reading your textbook.

4. **Work homework soon.**   Start reviewing and working homework as soon as possible to give yourself time to return to the problems you cannot do immediately. Do all of the homework.

5. **Study in addition to doing homework.**   Homework alone will not integrate and synthesize what you need to learn in your mind. Chapter 5 will give you specific study ideas and show you how to make note cards that will help you internalize and understand the material.

6. **Meet your resources.**   Introduce yourself to your teacher, available tutors, and fellow classmates so you have ready resources when you have questions or need company or motivation. Set up and meet with a study group. Chapter 6 will give you suggestions for how to use these resources.

7. **Stay the course.**   Be persistent. Don't quit. Go to class the whole semester even if you drop the course. Read the section "How to Make a Comeback!" at the end of this chapter.

8. **Keep a record.**   Record all your exam, quiz, and homework grades. Track your grade throughout the class.

Critical Time: The End of the Semester

Plan ahead for the end of the semester. It will be easier if you are proactive at the beginning of the semester and follow through with the activities that we have suggested. The end is a critical time and can make or break your grade in your math class. The last few weeks of the semester will be filled with final papers, the last test, review work, and the final exam. Do not add to this flurry of activity by getting behind on your math homework, expecting to make it up at the end. It won't happen. The last few weeks are when you might cut back on your work schedule outside of school, ask for additional help at home with chores, and assign additional hours to studying math.

Check the nearby list for a successful conclusion to your hard work all semester.

How to Make a Comeback!

Sometimes we are not prepared for a class or we make poor choices about homework early in the semester or we have personal issues that cause us to get behind. If that happens, don't quit even if you have to drop the course. Be persistent and go to class the whole semester. Here are three more stories from our tutors about their successful comebacks.

 SHORT-TERM GOALS FOR FLOW AT THE END OF THE SEMESTER

1. **Do math.**   Begin two weeks in advance of your final exam to work through your class notes and note cards using pencil and paper to work problems. Make a list of problems that you believe will be on the final.

2. **Master previous exams.**   Assemble all of your chapter exams from the semester. List problems that you missed. One by one, make certain that you now know that material and can work those problems.

3. **Ask for a final exam review sheet.**   Your teacher may have a review sheet specifically for the final exam. If he does, request a copy as soon as possible and begin working with paper and pencil.

4. **Meet with a study group.**   If you are not already in a working study group, form one to meet daily for the two weeks before the final. Focus on one chapter each day that you meet. Work problems from the class exams and problems from the end of the chapters. See Chapter 6 for suggestions about study groups.

5. **Surround yourself with math.**   Schedule yourself into the math tutoring center or the math study center or wherever other students will be working on math. Their energy toward math will energize you toward your math.

6. **Check the record.**   Talk to your teacher to be certain that you have completed and turned in all of the assignments. Compare your record of your grades with your teacher's record. Your teacher can help you estimate what score you will need on your final exam for the grade you wish.

7. **Build confidence through practice.**   Practice problems on your own several times a day to bolster your confidence. Review all of your note cards often. Work many, many problems and talk to yourself the entire time. Talking about the processes doubles the benefits.

8. **Learn test-taking strategies.**   Read Chapter 7, "Take Charge of Testing," to pick up new test-taking techniques for before, during, and after tests.

Alex's Breakthrough

"My worst math experience was getting a D in algebra. I thought it was the end of the world. It was very hard for me. I didn't know what I was doing wrong and what I was doing right. I got off to a good start with a B. Then everything went by so fast. I felt awful. I was O.K. with prealgebra and beginning algebra but, in second-semester algebra, I was always struggling the whole semester. I thought math was impossible for me. It lowered my self-esteem. I thought, why couldn't I do math? I saw everybody getting hundreds. When I would go to the Math Study Center, I would hear people use big mathematical terms. But now I'm just as good as they are at that. **My advice to struggling math students is 'Do not give up.' I thought I was going to give up because math had never been my favorite subject. I never gave up on it even though I struggled a lot."**

Isabella's Words of Wisdom

"I just walked through the feelings and the panic. When I would get tests back that were C or D grades, I would just say, 'Oh, well. It's just a test.' I knew I would be coming back the next class meeting. For me, there wasn't any option. My advice is to show up for class every day and show up on time. If you don't go to class, you miss stuff. If you're sitting there in class in an effort to pay attention—even if your mind is wandering—you're getting something. At least try to do the homework—at least try. If you get stuck, go on to the next problem. **Keep taking classes over again. Just keep taking them until you are ready to transfer or ready to graduate."**

How Jazmin Triumphed over Calculus

"The first time I took calculus II, I explained to the instructor that I was going to drop it a month before the end. He agreed that was the best idea, but I asked him if I could please attend the lectures because I was going to take it again that summer. I knew going through it once already was going to help me. It would make it easier the second time. When [my teacher] explained things in class and I didn't understand, I copied the notes down. I wrote all of the details, even things he would say. I knew that I was going to get it. Maybe not that day, maybe the next day or whenever, but I knew I was going to get it. I got excellent notes from that instructor. The best. When I took it again that summer, I used my first instructor's notes. I took them with me to the class the second semester with the new instructor. I didn't understand the new instructor and I couldn't get his notes or his style. I couldn't get anything. A lot of times I just had my first instructor's notes in front of me when the second instructor was lecturing. **When I couldn't understand him, I would look over the first instructor's notes. They were good notes. Very good. And I passed. I got it."**

ACT FOR SUCCESS | CHAPTER 4

1. With which math student (Enrique, Jazmin, Joel, Alex, Isabella, or Sarah) do you identify? Why? What is your favorite quotation?

2. Use the blank time schedule from the Appendix to make a sample weekly schedule for yourself. Use colored highlighters to block out times for Class, Sleep, Personal Care, Studying, Eating, Commuting, Family, Housekeeping Chores, and Work. Count the number of hours that you have assigned to math study for the week. If that number is less than two times the number of semester units for your math class, pencil in more time for math.

(continued)

MASTER MATH'S MYSTERIES

Introduction to Fractions: Adding and Subtracting

Why Fractions?

Fractions exist because the real world is not a whole number. Daily we experience half sandwiches, halfway home, half dollars, and doing half of the homework. We know about a quarter of a football game, quarter of a dollar, and quarterly payments.

But when we write $\frac{1}{2}$ sandwich, $\frac{1}{2}$ of the way home, $\frac{1}{2}$ of a dollar (.50 of a dollar), $\frac{1}{2}$ of the homework, $\frac{1}{4}$ of a football game, $\frac{1}{4}$ of a dollar (.25 of a dollar), and $\frac{1}{4}$ of a year, sometimes we panic and forget that we really know what that means.

What Is *One Half*?

Every fraction has a "whole" related to it. The "geometric couples" illustrated are different examples of "one half" and "one whole." $\frac{1}{2}$ of a sandwich is one of two equal parts of a whole sandwich.

The **number on top** of a fraction (called the *numerator*) tells "how many" or the "number." The **number on the bottom** (called the *denominator*) describes what we have.

For example: $\frac{3}{8}$ or three eighths means that we have three of the pieces we get from splitting some whole thing into eight equal parts. $\frac{3}{4}$, three fourths or three quarters, means that we have three of the pieces we get from splitting a whole football game, a whole dollar, or a whole year into four equal parts.

Operate on These Fractions

To add or subtract fractions, the denominators must be (and remain) the same because we need to add and subtract similar items. For example, $\frac{2}{4} + \frac{1}{4}$ is the same as saying two quarters of a game plus one more quarter of the game. That makes three quarters of the game. So $\frac{2}{4} + \frac{1}{4} = \frac{3}{4}$. Notice that the denominator stays the same. We merely added the numerators to see how many quarters we have.

Another example, $\frac{9}{10} - \frac{3}{10}$, is nine tenths minus three tenths, which is six tenths. Relating this problem to money, a dime is one tenth of a dollar. So $\frac{9}{10} - \frac{3}{10}$ becomes 9 dimes minus 3 dimes which is 6 dimes or $\frac{6}{10}$. Again, notice that the denominator does not change.

Check Out These Examples

Here are more examples of adding and subtracting fractions:

(a) $\dfrac{4}{7} - \dfrac{1}{7} = \dfrac{3}{7}$ **(b)** $\dfrac{11}{25} + \dfrac{6}{25} = \dfrac{17}{25}$

(c) $\dfrac{7}{8} - \dfrac{4}{8} = \dfrac{3}{8}$ **(d)** $\dfrac{4}{x} + \dfrac{12}{x} = \dfrac{16}{x}$

Notice that each time I add or subtract, the bottom (denominator) stays the same and I compute with the tops (numerators), which are the number of items. Even when I do not have a clear description of the item in example (d), I still add the tops and keep the bottom the same. (The "*x*" is just a dummy stand-in for an unknown number. Because there is an *x* in both denominators, I can still add the tops and keep the bottom the same.) If the bottoms (denominators) are different, I will need to use a different procedure that is discussed in "Master Math's Mysteries" in Chapter 5.

You can think about the preceding four examples in real-life terms too:

(a) "4 dogs minus 1 dog is 3 dogs" is similar to $\frac{4}{7} - \frac{1}{7} = \frac{3}{7}$.

(b) "11 apples plus 6 apples is 17 apples" is similar to $\frac{11}{25} + \frac{6}{25} = \frac{11}{25}$.

(c) "7 books minus 4 books is 3 books" is similar to $\frac{7}{8} - \frac{4}{8} = \frac{3}{8}$.

(d) "4 cars plus 12 cars is 16 cars" is similar to $\frac{4}{x} + \frac{12}{x} = \frac{16}{x}$.

Practice This Kind of Thinking

Work these examples. If you feel squiggly, return to the parts of the preceding explanations that helped you understand. Check your answers with the solutions in the Appendix.

1. $\dfrac{6}{10} - \dfrac{5}{10}$ **2.** $\dfrac{17}{37} - \dfrac{12}{37}$ **3.** $\dfrac{2}{8} + \dfrac{3}{8}$

4. $\dfrac{6}{12} + \dfrac{5}{12}$ **5.** $\dfrac{13}{x} - \dfrac{3}{x}$ **6.** $\dfrac{2}{12} + \dfrac{3}{12}$

Did you get feedback by checking your answers with those in the Appendix?

Working with Halves

Pie Story: Marie baked three pies for a family celebration. While she was out, her kids ate *half of a pie*. When she returned, she cried, "Oh, no! Only _____ pies left! That is not enough." How many pies did Marie say were left? Think about it.

Method 1: Imagine or draw or bake whole pies.

■ Marie had three pies.

■ One half a pie was devoured by her children.

(continued)

- That would be: three pies minus one half pie.
- They didn't touch two pies. Together they ate one half of the third pie, leaving one half.
- That would leave $2 + \frac{1}{2}$ pies. Marie cried: "Oh, no! Only $2\frac{1}{2}$ left! That is not enough."

Method 2: Imagine half pies.   A second way to add or subtract one half is to recognize how many "half pies" there are in the three whole pies and then subtract the one half.

Marie had three whole pies, which means she had six half pies because each whole pie contains two half pies. When her kids ate one half pie, she was left with five half pies. Of course, she would most likely recognize those five half pies as two whole and one half pies.

Three pies = six halves or

$$3 = \frac{6}{2} \text{ so } 3 - \frac{1}{2}$$

$$= \frac{6}{2} - \frac{1}{2}$$

$$= \text{six halves minus one half}$$

$$= \text{five halves}$$

$$= \text{two wholes and one half}$$

$$= 2\frac{1}{2} \text{ pies}$$

> **Note on how "five halves" becomes $2\frac{1}{2}$:** Changing five halves into two wholes and one half can be accomplished by using division. Notice that it works to think of five half pies as five divided by two, which is two with one left over, or two and one half. Here are three other examples to consider and visualize:
>
> $$\frac{9}{2} = 4\frac{1}{2} \qquad \frac{15}{2} = 7\frac{1}{2} \qquad \frac{24}{2} = 12$$

Pick Your Favorite.   Which of the two methods–thinking of whole pies or of half pies–do you prefer? Try to understand both methods because each one can increase your understanding of the problem in a different way. Practice adding and subtracting $\frac{1}{2}$. Sketch pie pictures to help you visualize. Talk yourself through to get these ideas into your ears. Make pies to get these problems into your skin and muscles. Think pies! (Answers are in the Appendix.)

7. $4 - \frac{1}{2}$ **8.** $\frac{1}{2} + 4$ **9.** $3 - 1\frac{1}{2}$

10. $4 - 2\frac{1}{2}$ **11.** $4 + 3\frac{1}{2}$ **12.** $4\frac{1}{2} - 2$

Consider that we could generalize our work here to thirds, fourths, fifths, sixths. . . . We could even say that Marie's kids ate $\frac{3}{100}$ of a pie. Not very likely, but it could happen. If they did, that means those kids cut one of Marie's scrumptious pies into 100 parts and ate three. That would leave two whole pies and 97 parts out of 100 in the third pie. So $3 - \frac{3}{100} = 2 + \frac{97}{100} = 2\frac{97}{100}$.

Imagine those sneaky kids and try these:

13. $4 - \frac{7}{8}$ **14.** $4 - \frac{2}{5}$

Student Success Story

Sarah Kershaw

Diagnosed with dyslexia as she worked her arithmetic problems in the second grade, Santa Ana College math student **Sarah Kershaw** did not learn to read until she was 12 years old. She first began to understand mathematics as an adult as she discussed theory with a math teacher. Recognizing that she learns best by working from generalizations down to the specifics, Sarah began devouring math books with that focus and taught herself the math she needed to know. **In her 30s, she became a university honors student because she discovered and mastered the school techniques that work for her.** She has learned how to approach learning when the symbols—whether letters or numbers—become squiggly as she examines them. At Santa Ana College, she skillfully tutored statistics and assisted with the development of classroom demonstration tools to explain statistics concepts. She trained math faculty and others at the college on working with students with learning disabilities.

Sarah likes the creativity in math. She sees math as a construct of the human mind that becomes a language used to explain the world. She loves that there is more than one path to arrive at an answer because she claims she is not very good at going the same way twice. She advises you to pursue your questions, saying, "I don't let the question go until I get an answer that I understand."

Bridge the Gaps with Study Skills

"Unless you try to do something beyond what you have already mastered, you will never grow."

RONALD E. OSBORN

In this chapter, you will find effective study skills that are basic to creating a successful, positive experience for learning math. These crucial skills are reading the textbook, taking notes, asking questions, studying math, working homework, and balancing your mind and body. The six Santa Ana College (SAC) math students who gave us their expertise in Chapter 4 will continue to share their experience of struggling with and conquering their math studies.

Crucial Skill #1: Reading a Math Textbook

Your math textbook is your first resource in learning math, but reading a math textbook is different from reading textbooks from other disciplines. In math, every single word counts and often means something different from everyday usage. Also, math ideas develop slowly and depend on the preceding material. Not understanding one example in a math book can hinder our understanding of many later concepts. These differences influence how to read a math book. Expect to read slowly and to return to passages over and over. Expect to use pencil and paper to understand parts of it. Expect that your

understanding will evolve—not occur immediately. Expect to remind yourself often, "I don't understand yet."

   Before the lecture, read the current section of the math textbook to just familiarize yourself with what is to be covered. **Read your book with pencil in hand. Look for unfamiliar vocabulary or concepts, listing them to make note cards later. As you work through the section in your textbook, stop and work each example on paper. This is how you actually learn the material.**

Example

Here are textbook notes from Chapter 3's "Master Math's Mysteries" using the **column method**. Notice that organizing material is on top, the vocabulary and concepts are on the left, and a sample problem with the work is on the right side.

Chapter 3 Master Math's Mysteries	Math 100
○ Main Idea: Order of Operations 　　Use PEMA 　　　work down 　P　Parentheses 　　　( ) or [ ] or { } 　E　Exponents ○ M　Multiply and Divide* 　A　Add and Subtract* 　*always left to right 　　if on the same level	$[8(2) - 4] + 9 - (5 - 4)^3$ $= [16 - 4] + 9 - (5 - 4)^3$　P with M $= [12] + 9 - (5 - 4)^3$　　P $= 12 + 9 - (1)^3$　　　P $= 12 + 9 - 1$　　　E $= 21 - 1$　　　A $= 20$　　　A
Main Idea: Important Vocabulary 　　Sum ○ 　Difference 　　Product 　　Quotient	Sum of 5 and 2 is $5 + 2$ or 7 Difference of 5 and 2 is $5 - 3$ or 3 Product of 5 and 2 is $5 \cdot 2$ or 10 Quotient of 5 and 2 is $5 \div 2$ or $\frac{5}{2}$

After the lecture, reread the entire section of the textbook *before* attempting the homework. As you rewrite your class notes, include notes from the book to stay organized by concept. Remember to put the chapter number and section number as a title on each page.

   Use this checklist to read your textbook with more understanding.

SHORT-TERM GOALS TO FLOW WITH READING MATH

1. **Write as you read.**　Read with a sharp pencil, eraser, highlighter, and scratch paper at hand. Write notes in your book so that when you review your text you remember the breakthroughs in understanding that you had previously and so that you know where you did not understand.

2. **Use your voice.**　Read the material and examples out loud to enhance your memory and understanding.

(continued)

3. **Give it time.**   Say "I do not understand yet" and keep going. Reread when necessary.

4. **Find the big picture.**   Pay attention to the headings and topics so that you have an overview of the content. Highlight only the extremely important information that you will want to revisit. Do not highlight everything.

5. **Work the examples.**   Work out all of the textbook examples on paper and compare them to the solutions in the text.

6. **Learn the vocabulary.**   Look up the definitions of any words of which you are unsure.

7. **Take your book.**   Carry your textbook to class and to your teacher with your questions. Write the answers to your questions in your textbook. You will want to review those questions again and again.

Isabella says:

"I read my math book with a pencil and I write in the margins. I write questions in the margins that I might ask in class. I break examples down [and] make little arrows."

Enrique explains how he reads his math book:

"Everything the book says is backed up by an example. I always have paper and pen at hand. I read and then I write. I've tried just reading it and I can't. My hand starts twitching and I say, "Give me that pen and paper. I've got to see how they do it." Wow! It makes a really big difference. By doing that example I remember the definition and know what it's about. Sometimes I hit a roadblock and I go back and do the example again."

Alex describes his breakthrough with math when he finally began reading his textbook:

"I started reading my math book when I repeated my algebra class, and it helped out a lot. I realized that the wording is very difficult and I just kept practicing. After a while, it seemed like my language. I was surprised. It's not easy but it started to work for me. Helping other students out with their math, I used the same language over and over, so reading the book became much easier. I do work examples through with paper and pencil. I write down the question. Even if it's a word problem, I write it down. Because when I write it down in my own handwriting, I understand it more as compared to just reading it. I always work out the problems. I write little notes in my book. I will write little marks like question marks. If I see a sentence with a vocabulary word, I'll write the definition . . . in the book so it's all in front of me."

Reading the Math Book Exercise

Read the section of the textbook your teacher has assigned for the next class. Take notes. Fold your paper vertically as shown in the column method earlier. Label the section number and topic. Place vocabulary, main ideas, and concepts on the left side and work example problems on the right side.

1. List all the math vocabulary.

2. After reading through the section once, write a statement that encompasses the main idea of the section.

3. As you read through the section a second time, work out each example on paper.

4. Paraphrase the important concepts that are covered.

5. Note any concepts that were difficult or will require extra reviewing. Usually I put a star in the margin of my notes. Write any questions you have for the teacher.

6. Look over your notes one more time before the beginning of class.

Crucial Skill #2: Note Taking

Participate actively in class by listening carefully, asking questions, and taking notes. Note taking is individual. Our SAC math students all have something different to suggest. Joel says, "I always take notes. I copy what's on the board and the examples. I ask questions sometimes to clarify something that I didn't know or understand." Meanwhile, Enrique reports, "For my notes, I have a separate notebook for each class. I don't mix them. I keep everything all in order." Alex explains how he overcomes his lack of understanding in class:

> "Usually I copy what the instructor writes on the board and I write little notes to myself. I also do different colors. I write questions like, 'Why do we do this?' If I am embarrassed to ask in class, I write a little note and try to figure it out by myself or ask somebody. Sometimes the homework answers it. I always write an explanation like, 'We can't do this because of a restriction.'"

Isabella and then Jazmin attest to the importance of listening as well as copying what the teacher says. Isabella describes one technique for making certain her notes are accurate:

> "Before returning to school, I could never imagine that anyone could take notes in a math class. Now I take notes really fast. I copy down everything the teacher writes on the board. But I also listen. This semester in statistics, I actually changed things. I listened more to what the instructor was saying. I listened for his key ideas and wrote them down in words rather than in numbers. Sometimes I would copy off other people's notes—other people that take good notes—in case I missed something."

Jazmin describes another technique for clarifying her notes:

> "I need to get as much information as I can. As the instructor writes, I copy every single little thing, but I have to watch. If the instructor is talking and writing, I listen and copy later. I need to hear him too. At the university, my instructors go really fast so I take a tape recorder. I still try to take notes. If I have key words, I can remember the whole story to it. But if I don't have key words or don't understand something then I'll listen to the tapes."

These successful math students have given you excellent strategies for taking notes:

- Copy what's on the board and the examples with explanatory notes.
- Listen carefully and write down summaries of what the teacher says.
- Ask questions for clarification.
- Keep a separate notebook for each class and keep notes in order.
- Borrow good notes from other students.
- Tape-record class to fill in details after class.

How to Take Notes

As the students have said, note taking is a balancing act between listening, watching, and writing important facts. To take notes in class, include the following five things as you write:

1. Organizing information: course name, date, chapter number, section number, and main concept.
2. Concepts, vocabulary with definitions, and notes that the teacher writes on the board.
3. Problems used as examples.
4. Solutions for problems with an explanation of the steps.
5. Summaries of what the teacher says.

I use the column method already introduced for note taking whether I am reading a textbook or note taking in class. I place a line or fold in my paper at a little less than halfway as shown in the illustration. I leave space for organizing information at the top; then I put general notes and vocabulary on the left side and examples on the right side. Here is an example of a two-column method for note taking in class.

Section 2.1 Order of Operations	Basic Math October 1, 2012

○ 2.1 Order of Operations
    Use PEMA and work down
    P Parentheses
○   ( ) or [ ] or { }
    E Exponents
    M Multiply and Divide*
    A Add and Subtract*
○ *Always work left to right
      if on the same level
	Example #1:
	$28 \div 4 - 3 + 4(5 - 2)^2$
	$28 \div 4 - 3 + 4(3)^2$ P
	$28 \div 4 - 3 + 4(9)$ E
	$7 - 3 + 4(9)$ M
	(left to right)
	$7 - 3 + 36$ M
	$4 + 36$ A
	40 A

Note-Taking Exercise

Practice using a column method for notes as shown in the checklist. Place a line or fold vertically in your paper at a little less than halfway. Leave space for organizing information at the top; then put general notes and vocabulary on the left side and examples on the right side. Use this method for notes in one of your classes today.

Use this checklist in addition to the suggestions already given to you by the SAC math students so that your notes are thorough and useful for you.

MORE TIPS TO MASTER NOTE TAKING

1. **Get set for class.** Be prepared to begin taking notes the moment class begins. Often what is said during the first few minutes sets the stage for the class and gives an important overview of the material. This is worth noting and remembering.

2. **Don't miss the conclusion of class.** Keep taking notes until the moment class ends. Often what is said during the last few minutes summarizes the day's class and sets the stage for the next class. This is worth noting and remembering.

3. **Keep your textbook open.** Turning to the section of the text being discussed will help you spell words and see what the teacher discusses.

4. **Pick a partner.** Sit by another proactive, focused student and compare notes. One of you may catch what the other misses.

5. **Use shorthand.**   Develop your own shorthand style for abbreviations. If a math word such as *equation* is used repeatedly, you could put "eqn" in parentheses after "equation" and then abbreviate after that.

6. **Organize along the way.**   Always put the name of the class and date on top of your paper. Put the chapter and section number as a title on each page. Use numbers 1, 2, 3, . . . or a, b, c, . . . or A, B, C, . . . to list subtopics. Separate examples with lines or boxes or circles or empty space. Note teacher comments such as "on the next test," "important," "tricky," or "remember this."

7. **Mark vocabulary.**   Highlight or mark with red ink all important vocabulary words. Leave space for definitions and fill those in.

8. **Make spare minutes count.**   Use dead class time while you are waiting to make your notes more legible, complete, and understandable. Identify main topics and subtopics and write them at the appropriate places in your notes.

9. **Pick your style.**   Make your notes in a format that will be clear for your use as you reread and quiz yourself later. Your notes are a clue to the way that your brain organized the material. Reworking your notes regenerates and strengthens the brain connections you are making with the subject. **You take notes quickly during lectures, and it is when you review your notes and rework them that you actually learn the material.**

10. **Complete your notes.**   Add additional explanations soon after class when your memory is fresh. Compare your notes to others after class if you believe yours are not accurate or complete.

11. **Transcribe your notes.**   Organizing your notes will make review easier later on. Rewrite them soon after class and make classroom examples stand out with explanations so that they are easy to find. If you do not wish to rewrite your notes, use a highlighter and a different colored pen to circle, box, and highlight important notes and problems.

12. **Identify test questions.**   Find classroom examples to write a list of predicted problems for the next test. Use these problems as a dress rehearsal test for your next exam.

Crucial Skill #3: Asking Questions

Ask questions of yourself, your peers, your teachers, your textbook, and your tutors. Asking those questions is essential to learning mathematics. Mark them in your textbook, notes, or your homework. No one else will know or have the same questions you have. Asking questions is a fundamental right in learning math and, frankly, there is no other way to learn. One small question can mean the difference between a "Eureka, I've got it" experience and an "I don't get it at all" experience. Which do you prefer?

TRY THESE POWERFUL QUESTIONS

What if we tried this?   When would this process work?

What caused this step?   Tell me about this piece.

Where would this happen?   How do I recognize the difference?

How could this happen?   What else might work here?

Sarah's advice and comments are exactly what countless students have told me about their experience, but few of them have been as brave as Sarah. Consider adopting her attitude toward questions as a role model for yourself. She says:

> "I ask the stupidest question if I need to. I don't let the question go until I get an answer that I understand. It [can be] embarrassing. [Sometimes] the people behind start whispering because they get frustrated with me. But that's only one or two people. I have people who come up after class who say, 'Thank you. I didn't get that either.'"

The following checklist will help you become more comfortable asking the questions you need to ask whether you bravely ask them in class like Sarah or ask them outside class of the instructor, tutor, or another student.

ACTIONS TO MAXIMIZE QUESTIONS

1. **Prepare a prioritized list.**   List your questions beforehand so you remember them. Number them from the most to the least important. Copy the problem that the question came from.

2. **Ignore others in class.**   Sit in the front of the classroom so you don't see any other student's reaction to your questions. Their reactions don't matter but sometimes they stop us from asking what we need to ask.

3. **Take a chance.**   Be brave. Even if you've never asked questions in math before, try new behavior. The powerful questions listed earlier can help you craft outstanding questions. In my opinion, there is no way to avoid asking questions if you want to learn.

4. **Enlist support.**   Get a partner to assist you in asking your question. Go together to see the teacher or just say in class, "Sally and I were wondering," then ask.

5. **Ask often.**   The more you ask, the more comfortable you feel. You might ask a question to which you already know the answer for practice.

6. **Show your work in public.**   Write questions on the board before class begins. On the classroom sideboard, work the problem out as far as you got. Then the teacher can specifically answer your question.

7. **Be O.K. with no answer yet.**   Expect that your questions will not be answered every time you ask. That does not mean the question was not a good one. There are time constraints in class. Ask again later.

Crucial Skill #4: Studying

Successful math students study as well as do homework. What does it mean to study? Most students don't know. Alex says:

> "I don't think just doing the homework is enough. I do the homework and make sure that I understand the concepts. Once I complete the homework, I ask myself, *"Do I really understand this?"* If I don't understand it, I've probably just done the mechanics by following along with an example from my notes or the book. That helps me get the problem but it doesn't help me out in the long run unless I spend time understanding it."

Isabella tells us her successful study techniques:

> "When I study, I *recopy the notes*. I draw pictures, make arrows, and write things down in *steps*. I make an *introductory sentence* about what we're trying to do here as the teacher has said it. I will write, 'We're trying to find the average' and then put the steps to do it. I put a big number one and circle it. I *write out the first step in words* and then I write out the formula way to do it. Sometimes my numerical illustrations have arrows or captions or bubbles to explain what I am seeing.

> "I *draw myself a little box titled 'Skills Required.'* Then I make a list of the skills required such as being able to take an average. That's good for studying. Sometimes to prepare for a test, I'll write down 'Skills Required for Chapter 8.' Then I'll go through and *make a list of all the things I've learned to do*. I even have detailed notes with little pictures that I drew of the buttons on my calculator."

Use this checklist to develop specific activities for studying beyond working homework problems:

 ## SKILLS TO FLOW WITH STUDYING MATH

1. **Be self-sufficient at first.**   Read the textbook. Work the examples from the textbook over and over until you can work them without referring to the solutions in the text.

2. **Rework class notes.**   Recopy and organize your class notes.

3. **Make note cards.**   Use index cards to make two kinds of note cards to organize the material and to quiz yourself during your spare time. See the explanation of note cards and note-card activity later.

4. **Study with a group.**   Meet with a study group to outline what you believe to be the content of the next exam. Collect one or two problems for each concept, and then work them together on a whiteboard describing the processes out loud as you work. Take turns explaining the problems to each other. Chapter 6 will help you further with study groups.

5. **Take practice tests.**   Ask if your instructor makes old exams or a review sheet available to students. If not, write a practice exam for your next test. Use the review section at the end of each chapter to choose two or three problems for each section of the chapter or choose two or three odd problems from each section of the homework. Make certain your problems cover each concept mentioned in your notes. Write an answer key for your exam. Then take the test and check yourself. Better to mess up now than on the real thing.

6. **Discuss problems.**   After correcting your practice tests, work these problems with another student and discuss them. You will be surprised how much more you remember when you discuss math out loud with someone else than when you work quietly alone.

7. **Focus on vocabulary.**   Make a list of all the vocabulary words from the chapter with their definitions and examples illustrating the definitions.

8. **Work out loud alone.**   Save time and learn the vocabulary easily by reading all homework problems aloud and verbalizing the processes using the correct vocabulary as you work.

9. **Use the supplements.**   Some textbooks come with ancillary materials such as study guides or online quizzes. Use them to test yourself.

Making Note Cards

There are two main types of note cards—Q&A cards (with questions on one side and answers on the other) and informational cards. For either type of card, put the section number in small writing on a bottom corner in case you need to refer back to the book for more information. You can use color to help organize your cards. Punch holes in an upper corner and tie the cards loosely together for easy review anywhere.

Q&A Cards.   As you review your lecture notes and the textbook, make note cards for new vocabulary, symbols, and sample problems. Place the new vocabulary or symbol on one side of the card with the definition on the other side. Place the sample problem on one side with the work on the other side.

FRONT SIDE OF CARD   **BACK SIDE OF CARD**

Example 1: Vocabulary

PRODUCT Chapter 3	means "to MULTIPLY"

Example 2: Symbol

4 · 6 Chapter 2	means "four times six"

Example 3: Sample Problem

Use compatible numbers to add: 366 + 647 Chapter 1	366 + 647 = 6 + 360 + 640 + 7 = 1,000 + 13 = 1,013

Informational Cards.   Use the cards to note multiple steps or lists of information. The following example showing the order of operations with an example on the other side is an information card. All of the lists of short-term goals or strategies in this book are also examples of information cards.

Example

FRONT OF CARD   **BACK OF CARD**

P Parentheses
*   ( ) or [ ] or { }*
E Exponents
M Multiply and Divide
A Add and Subtract
*   always left to right,*
*   if on the same level*

$$28 \div 4 + 4(5-2)^2$$
$$28 \div 4 + 4(3)^2 \qquad P$$
$$28 \div 4 + 4(9) \qquad E$$
$$7 + 4(9) \qquad M$$
$$7 + 36 \qquad M$$
$$43 \qquad A$$

Note-Card Exercise

Read the section from your textbook that will be covered in the next class. Make note cards from the section's "Master Math's Mysteries."

1. Make Q&A note cards for all math vocabulary, symbols, and example problems.
2. Make informational cards of concepts or steps.
3. Punch holes in the cards and tie them together.

Crucial Skill #5: Math Homework

Research studies show that students who work homework pass at a much higher rate than those who do not do homework. In my own research of nearly 1,000 algebra students at my college, one of the two best predictors of passing the intermediate algebra class was completing homework. Jazmin knows herself well and has discovered what gets her homework completed:

> "I make myself *go to the math tutoring center right after my class* and not go home. I know that if I go home, I won't do it. It is easier for me to study here on campus. I try to *do my homework right after class*. When I took a class on Monday and Wednesday, I would say I would do my homework on Thursday or Friday and that didn't work."

Isabella describes her winning strategies:

> "I have to practice math more than other subjects. I have to do the homework. My notes from math class are useful to me. I can actually *go back to my notes* when I can't do the homework and I can see what I'm supposed to do. I do get to my homework during the same day and I get it done and get it done well. Sometimes I want to start and can't wait—especially in statistics. He gave us a lot of worksheets in class. I would go back and finish those worksheets because basically I had just copied down some things without understanding them. I would *not start on the homework but rather on the worksheets* so I could go over what we had done in class on that day in the way that we did it."

Be on the winning team and use the following checklist to set short-term goals for your after-class work.

FLOW WITH HOMEWORK

1. **Know the assignment.**   Know it exactly. Be meticulous daily about writing assignments neatly and carefully in a pre-designated place in your notebook. If there is an assignment sheet, make several copies so that you are never without it. Place one in your textbook and one in your notebook.

2. **Begin right away.**   Start homework soon after class. Your brain chemistry is still focused on math and you can take advantage of dendrite networks (brain connections) that are newly formed. Studying immediately solidifies those dendrite networks so that you understand better and remember longer. **If you study immediately after class, you will have to spend less time studying math in the long run because you are using your brain more efficiently.** Studying soon also gives you time to mull over your questions and get them answered before moving on to new material at the next class.

3. **Use class examples to warm up.**   Rework and rewrite your notes and make certain that they are written clearly and make sense. Work the examples done by the teacher in class several times to warm up and learn the concepts before you begin homework problems.

4. **Work textbook examples for a head start.**   The homework problems in a math text come in the same order as the textbook explanation and examples. Read and work the examples in the textbook to solidify the concepts and processes before you work homework problems. As you work homework, look back to the section in the text where those problems are explained.

5. **Use paper.**   Be prepared to use lots of paper. Give each problem plenty of space on the paper so that you can write legibly and comfortably. Leave space between problems to write questions and to frame your work for later review. If you get stuck, begin again on a large clean paper so that your mind is cleared of what you already tried. Mathematicians use lots of scratch paper and begin again often!

6. **Line up numbers with graph paper.**   Work homework problems on graph paper to keep columns straight. I use "Engineers Computation Pad" graph paper. Graph paper is highly recommended and works well for dyslexic students. It helps keep the numbers and problems from getting squiggly.

7. **Use your support team.**   Arrange for visits to your resources by making appointments or dropping in during your teacher's office hours. Plan ahead for working with your teacher, tutor, or study group so that you know you have a planned resource for those inevitable questions that we all have when we are learning and working new material. Read Chapter 6 for more ideas for using resources.

Alex's strategies for homework indicate his determination to answer his own questions, if possible:

> "To do homework, I *look at the examples*. First I always *try to answer my own questions by looking at the book or by looking at the notes*. I really don't care how long it takes me to answer. I just want to see if I can answer it myself before I ask anybody else. If I let somebody else answer it, I see what's right but I never put the thought into finding the answer. I think that *if I look at the examples I can learn on my own*. That helps me remember more."

Joel shares his winning homework process:

> "I identify what kinds of problems there are and know how to do each kind. I recommend *knowing the differences between the kinds of problems*. I always do all my homework. A homework problem could be on my test."

Crucial Skill #6: Using the Mind/Body Connection

Your mind and your body are interconnected. Your thoughts are created by electrical and chemical reactions throughout your body. Using common sense about your food intake, water intake, sleep, and exercise affects those electrical and chemical processes positively and ultimately that influences whether or not you understand and remember your math. **A tired, dehydrated, out-of-shape, and malnourished body will have difficulty learning math.** Keep your body in tip-top shape.

Use the following checklist to keep your mind working to the max.

TIPS TO MAXIMIZE YOUR MIND/BODY CONNECTION

1. **Exercise.**   Contribute to your body's well-being with exercise such as walking, yoga, calisthenics, riding a bicycle, or running. When you are taking a math class, continue your regular exercise routine. Do not introduce a strenuous routine you are not used to. If you don't exercise, just walk and stretch a little more each day as you breathe deeply. Good blood circulation keeps you sharper.

2. **Watch what you eat.**   Regular nutritious and balanced meals keep you healthy and contribute to keeping a clear mind. The kinds of food that work best for you will differ from what works best for other people. Notice which foods and what amounts of food keep you the most alert. Fruits and vegetables are always good choices, but a balanced diet is essential.

3. **Breathe.**   Breathe deeply when you are working on math in or out of the classroom. The bloodstream carries oxygen to nourish your brain. Not only will breathing deeply oxygenate your blood, but it also will relax you and help you focus.

4. **Monitor caffeine.**   The right amount is an individual issue. If you need your coffee, drink it. If caffeine makes you hyper, skip it.

5. **Stay hydrated.**   Drink water to keep your body well hydrated. Dehydration causes fatigue and lack of clarity. Bring a water bottle to class.

6. **Sleep.**   Get the right amount of regular sleep for you. Lack of sleep has been shown to impair memory. Sometimes you can't do a math problem just because you are tired.

ACT FOR SUCCESS | CHAPTER 5

1. Reread all of the suggestions for mastering note taking. Write 5 short-term goal activities for yourself to improve your math note-taking skills. Use them in class and afterward assess how well they worked.

2. Use the column method for note taking in class. After class, reread the note-taking section in this book and write down three things you can do next time to improve your notes.

(continued)

3. *Use the note card activity to make both Q&A note cards and organizational note cards for a chapter in your math textbook. If you are not currently in a math class, use Chapter 3, 4, or 5 in this book.*

4. *Read your math textbook using the procedure outlined in this chapter, using pencil and paper to make notes.*

5. *Write a page describing how to study math successfully. Include the SAC student tips or any other success strategies found in this chapter.*

MASTER MATH'S MYSTERIES

More on Adding and Subtracting Fractions

In this Master Math's Mysteries, we discuss adding and subtracting when the denominators are not the same. This discussion is just a beginning for this topic, but it will, I hope, give you a foundation for continued work in a regular math textbook.

It's All in the Game

Any sports fan knows that adding the first half of a game and one more quarter brings the game to the end of the third quarter. In mathematical symbols, that means $\frac{1}{2} + \frac{1}{4} = \frac{3}{4}$. There is really no way to use the numbers given in $\frac{1}{2} + \frac{1}{4}$ to get $\frac{3}{4}$ *except* to recognize that the first half is actually two quarters or $\frac{1}{2} = \frac{2}{4}$. Then a middle step allows us to add like this: $\frac{1}{2} + \frac{1}{4} = \frac{2}{4} + \frac{1}{4} = \frac{3}{4}$. When the bottoms or denominators become the same, we use the method for adding shown in Chapter 4. Recall that the method was to add the tops (numerators) and keep the bottom (denominator) the same.

Here Is the Bottom Line

We must get the same denominator on fractions in order to add or subtract them. And it is simplest to use the smallest numbers possible, so we call the "best bottom number" the "*Least Common Denominator,*" or LCD for short.

In our sports example, the 4 was the LCD for 2 and 4. To absorb this, you may need to draw a picture or make a three-dimensional model using clay. You might want to split a pumpkin pie or pizza into four equal parts to see that together two of those four parts can either be named two fourths or one half. This is the same idea that two quarters of a football game is one half. An excellent model for adding and subtracting fractions can even be made from an empty egg carton.

How to Turn an Egg Carton into a Fraction Calculator

A carton from a dozen eggs is an easy model to make and use for fractions. The egg carton has 12 pockets for eggs–12 pockets for 1 *whole* dozen. Each pocket is one twelfth or $\frac{1}{12}$ of the *whole* carton.

Looking at your egg carton, ask yourself these questions:

■ How many pockets would be in one third of the carton? Splitting the carton into three equal parts puts four pockets in each part. So four pockets or four twelfths would be one third, and $\frac{4}{12} = \frac{1}{3}$.

■ How many pockets would be in two thirds of the carton? The answer is eight pockets or eight twelfths, so $\frac{8}{12} = \frac{2}{3}$.

■ How many pockets would be in one fourth of the carton? Splitting the 12 pockets into four equal parts puts three pockets in each quarter. So three pockets or three twelfths makes one fourth. Or $\frac{3}{12} = \frac{1}{4}$.

■ How many pockets would be in three fourths? Three fourths = nine pockets or nine twelfths.

If you have extra egg cartons, try cutting them into halves, thirds, fourths, and sixths.

1. How many pockets would be in one sixth? _____

   Two sixths? _____ Three sixths? _____ Four sixths? _____ Five sixths? _____

   (continued)

Put Your Egg Carton to Work.   Use the Egg Carton Calculator to add and subtract.

a. $\dfrac{1}{3} + \dfrac{1}{12} = \dfrac{4}{12} + \dfrac{1}{12} = \dfrac{5}{12}$

One third of the carton plus one twelfth of the carton is four pockets plus one pocket, which is five pockets or five twelfths.

b. $\dfrac{2}{3} - \dfrac{1}{4} = \dfrac{8}{12} - \dfrac{3}{12} = \dfrac{5}{12}$

Two thirds of the carton minus one fourth of the carton is eight pockets minus three pockets. The answer is five pockets or five twelfths.

Notice that *every fraction had to be renamed with a common bottom number (denominator) before adding or subtracting.*

Practice Makes Perfect.   Use your Egg Carton Calculator again.

2. $\dfrac{1}{3} + \dfrac{1}{4}$ 3. $\dfrac{3}{4} + \dfrac{1}{12}$ 4. $\dfrac{1}{6} + \dfrac{1}{3}$ 5. $\dfrac{1}{6} + \dfrac{1}{4}$

6. $\dfrac{2}{3} - \dfrac{1}{12}$ 7. $\dfrac{3}{4} - \dfrac{1}{3}$ 8. $\dfrac{5}{6} - \dfrac{1}{4}$ 9. $\dfrac{1}{3} - \dfrac{1}{4}$

Create Your Own.   Make up problems and check them with your Egg Carton Calculator. (Use only twelfths, sixths, fourths, thirds, and halves in your problems. The egg carton does have its limits.) Make up problems to add or subtract three fractions; for example:

10. $\dfrac{1}{6} + \dfrac{1}{3} + \dfrac{1}{4}$

Look Back–Comprehensive Review for Your Practice

Simplify:

1. $5 \cdot 40$ 2. $1{,}110 + 1{,}890$

Fill in the blank:

3. The sum of 9 and 4 is _____.

4. The difference of 12 and 5 is _____.

Perform the indicated operation:

5. $\dfrac{3}{4} + \dfrac{1}{4}$ 6. $31 + 75$

7. $24 - 2(16 \div 2) - 5$ 8. $6 - \dfrac{1}{2}$

Perform the indicated operation:

9. $\dfrac{5}{7} - \dfrac{2}{7}$ 10. $35 \cdot 10$

Student Success Story

The Tortoise and the Hares

A colleague told me a story about a woman in one of his graduate math courses who consistently asked questions that seemed very basic, almost too simple. My colleague and the other students began to roll their eyes and secretly smile whenever she raised her hand. They even quietly whispered to each other, wondering how she got into such a high-level math course. When final grades were revealed, there was only one outstanding score on the final exam and in the course. It was hers.

Use All Your Resources

"Nothing is troublesome that we do willingly."

THOMAS JEFFERSON

Resources enrich and enable learning. Your resources are your textbook, classroom, study environment, teacher, study groups, and tutors. This chapter will show you how to facilitate and interact with these valuable resources. Other resources include fellow students, other math teachers, other tutors, math students standing in the hallways, math tutoring centers, online tutoring, and other textbooks from the library. Getting acquainted with resources when you are calm and relaxed helps you more easily find them later on when you're stressed and urgently need help. These resources can be your life raft when you feel like you're going under.

How do resources enrich and enable learning? They answer our questions, and these questions enable us to learn. Learning is a repetitive cycle in which, first, we sense something is new and different; second, we test our understanding of it; and then third, we get feedback to refine our understanding. This learning cycle repeats quickly and continuously during class and study time. It is our questions (What if? How could? When would? What was? How do? Why does?) that keep the cycle moving.

Questions unlock the unknown. No one else will know your questions or even have the same questions you have. Your questions are based on your individual experience and physical brain structure. Questions clarify your understanding because they wire the correct pathways (dendrite networks) in your mind—pathways you use over and over to learn deeper and more difficult material. Math student Sarah Kershaw says:

> "We assume that what's familiar is simple when it really isn't. It took thousands of years to develop [math] and bring it all together. It didn't just pop into Newton's head one day. Each bit builds on the other."

How to Use Your Valuable Resources

Resource #1: Your Math Textbook

Your textbook is your first and best resource. Learn how to use it well and go there first when you have a question. Digging on your own before you go to someone else will help you understand if and when you do need to ask another person. Review "Crucial Skill #1: Read a Math Textbook" in Chapter 5 for how to use your book.

Resource #2: Classroom Instruction

Choose to participate actively in the classroom environment. To maximize your time, review Chapter 4 sections "Critical Time: Daily Before Class" and "Critical Time: During Class." Getting to know your teacher and other students one on one increases your comfort level by giving you allies.

"I now ask small questions in my math class. At times I'll get the general concept of what's being taught, and, before, I would never ask those little questions that would help me to clarify the bigger picture."

Andrea, algebra student

Unfortunately, some math classrooms have an environment of fear. Neuroscientists know that threat causes "emotional flooding" in the brain and temporarily renders the brain biologically and chemically unable to learn.

You can recognize the difference between a safe learning situation and a threatening one by how questions are handled.

■ In a safe learning environment, all questions are acceptable but time constraints are a reality. In classrooms, students often have to get their answers outside lecture because of limited class time. As a student, ask all the questions you have and let the teacher judge the time constraints. The worst thing that can happen is that your question won't get answered the first time you ask it. Eventually you will get an answer.

■ In a threatening environment, questions may be judged unworthy. This is *not* acceptable. There are no unworthy questions—no bad questions and no stupid questions. Questions are requests for information that students do not have and do not know. If you do not know something, isn't that the reason you are asking? When you don't know something, you don't know it.

Students have a right to ask all their questions. At the same time, the teacher or tutor has a right *not* to answer all of them because of time limitations. Judging students' questions is no one's right.

Even as you concentrate, you still won't catch everything. When the material is mostly new, much of what is said goes by you. That is the human condition. That is normal. Your brain needs processing time to integrate information with previous ideas. Asking questions helps you to sort, process, clarify, integrate, and understand.

When questions are not asked in a classroom, that doesn't mean everyone understands. It can mean that people are so confused they don't know where to start. If you ask your question, you might just clarify the session for others as well as for yourself.

SHORT-TERM GOALS FOR THE BEST STUDY ENVIRONMENT

1. **Find a good spot.**   Find a place that is free from distractions by your friends and your family. Set aside enough time for yourself. This could be in the school library, in an empty classroom, at a tutoring center, in your bedroom, at the public library, or even in the bathroom. Some students study at coffee shops or bookstores.

2. **Control sound.**   Arrange for pleasing background sounds or wear earplugs. You could listen to your own music with a headset.

3. **Keep supplies ready.**   Surround yourself with all of your materials–notes, textbook, syllabus, schedule, previous homework, teacher worksheets, previous chapter tests, pencils, erasers, and lots of paper.

4. **Make it pleasing.**   If you are fortunate to have your own desk or study table in your home, keep it inviting and prepared for your work at all times. Put an attractive and relaxing picture within view to stare at for short breaks from your work.

5. **Match your style.**   Review Chapter 3, "Learn How You Learn," to personalize your study environment to fit your unique learning styles.

Resource #3: Personal Study Environment

Study environments are individual but extremely important. You do not need your own private office to do your best work. You can adapt to other environments as long as you know exactly what you need. Witness how much work gets done in airports and on airplanes where people are surrounded by noise and other people. Use this checklist as you develop the best study environment for you.

Resource #4: Teacher

Communicate with your instructor. When a student expresses goodwill, shows a desire to learn, studies, puts energy into homework, and comes to class, teachers are delighted. When a student interacts with them, teachers are often willing to go far beyond their contract requirements to assist.

Math teachers want to reach their students. They become math teachers because they care about students, they care about learning, and they care about mathematics. They often blame themselves when students are not doing well. Occasionally teachers become defensive and feel that questions criticize them. A few others have forgotten how difficult it is to learn math. Many teachers have learning modes and combinations of intelligences that are different from their students'. Most math teachers work very hard at their jobs and wish the best for students. They do become frustrated when it *appears* as if students don't care about their own studies. This is why it is important for you to communicate concerns and problems that you are having with your math work. Teachers cannot read your mind. You will want your teacher to know about your goodwill and your efforts—not to expect a higher grade than reflected by your skill level but to keep your teacher as a supportive resource to plan your success with math.

As chair of my college math department, I received phone calls and visits from students complaining about or frustrated with their teachers. Usually, with goodwill and patience on everyone's part, I arranged for the student to talk to the teacher. Often this worked wonders. I received many follow-up phone calls from students reporting how understanding their teacher had been and that class was going better.

By communicating with your instructors, you give them necessary feedback and you help them understand your individual issues. After a one-on-one talk, both you and your instructor see each other's points of view a little better. It may be difficult for your instructor to listen to what you have to say, but it will be helpful, and almost all of them will be extremely grateful. If you need some support to speak with an instructor, department chairs, deans, and counselors can give you ideas or might accompany you to assist the discussion.

If you have gone the extra mile to communicate with your instructor and it hasn't helped, attend class, tape the lectures, and get a tutor to help you. **Do not let anything stand in your way of learning math.** The math department chair, dean, or college counselors can help you find alternative resources.

"Fight to have a good teacher. Lobby for people who understand mathematical thinking."

Judy Schaftenaar, Ph.D., educator and administrator

"There are ways to treat teachers so that they will respond to you."

Elinor Peace Bailey, artist, author, teacher

Use this checklist to guide your interactions with your teachers:

HOW TO GET SUPPORT FROM TEACHERS

1. **"Play the game."**   Smile when the teacher reads your name on the roll. Act interested. Don't monopolize classroom air time but participate in discussions with answers and questions.

2. **Check in.**   Stop by the teacher's office or desk after class and introduce yourself. Share how this class will help your hopes and dreams. Share any concerns you have about your performance. Let your teacher know you as a student.

3. **Be proactive.**   Come prepared. Keep the teacher informed about your progress. Ask questions you have written in your notes.

4. **Show kindness.**   Treat the teacher respectfully as a fellow human being. Smile and greet the teacher at the start of class.

5. **Communicate.**   Inform the teacher if a personal event hinders your studies. Do not just disappear one week and reappear the next. Do not use these events as excuses. Learn the rules of the school for absences and incompletes so you know your options. Do not expect the teacher to give you a good grade if you have not fulfilled the requirements.

6. **Participate.**   Volunteer answers to the teacher's questions. Help set up or take down the classroom so the teacher has time to interact with students before or after class.

7. **Know the boundaries.**   Read the course syllabus and refer to it often so that you know the ground rules and expectations of the teacher in your course. Do not expect allowances that other students would not receive.

Artist, author, and teacher Elinor Peace Bailey recalls her favorite teacher:

> "I talked to some of my fellow students and they found [the teacher] dry and uninteresting. But I went to him after class. When I had a [subject] that interested me, I would bug him until he would sort of 'pour forth' the enormous knowledge that he had. I was taught because I demanded to be taught—not in a way to accuse him or to put him on the defensive but in a way that made him feel sympathetic and cheered by my interest. I learned a good deal from him."

SAC math student Alex Solano says:

> "I remember when I went to my instructor's office the first time to ask a few questions, I was kind of embarrassed but I said, 'I have to do this or I won't understand.' Probably for someone to be comfortable with an instructor they need to go to their office hours. In class you can only talk to them for a little time before class or during class or after class. But in office hours, they're there. I go to office hours. The instructors are sitting down and they're there to listen to you. [Now] I get along with the math department and all my instructors so I feel comfortable talking to them. I get comfortable with someone when I see them a lot."

Resource #5: Study Groups

"It's always easier if you're doing something hard if you have other people to do it with."

Mick Cornett, Oklahoma City mayor (*Time*, January 21, 2008, p. 16)

Small-group study works well. We become acquainted with our classmates while we interact actively with the subject matter. As we talk, write, and do problems together in a group, we have the benefit of other minds as well as our own. We don't get stuck as often because we can frequently answer each other's questions. It is more fun struggling together.

Someone has to organize a study group or facilitate a support system within the classroom. Otherwise those things don't happen. Why not you? How would you do this? The following are suggestions for initiating a study group. Try one or all three of them.

- Request that the teacher announce the forming of a study group and tell interested students to see you.
- Talk to a number of students and suggest forming a group. Let those interested help you plan.
- Announce the group yourself, giving time and place. The library, an empty classroom, the math study center, and the cafeteria are possible locations.

To make the group effective, develop an open policy that allows all questions. Encourage a supportive attitude and a positive tone. Invite others from your class to join your group.

Use this checklist to set up a winning structure for your group. Make a copy for each group member.

SHORT-TERM GOALS FOR SUCCESSFUL STUDY GROUPS

1. **Set it up.**   Exchange names and contact information and plan a meeting place and schedule.

2. **Combine focus and fun.**   Have fun and laugh and *never forget* that your goal is to learn math together. Make corny math jokes as you work problems and answer questions. Keep the time congenial *and moving forward*.

3. **Commit to the group.**   Expect each group participant to do the homework and to bring questions to the group meeting. All of a student's math study cannot be done in a group. Each person needs to be responsible to add to the group and push everyone's understandings further. This cannot be done without time and work.

4. **Work together.**   Discuss class notes and homework questions. The rules for a discussion are: You talk, you listen, you talk, you listen, ...

5. **Plan together.**   Make an overview of what the course is covering at the moment. Prepare a common list of expected problems on the next test. Make a group plan to obtain and share as much advance information about the next exam as possible. Do this by taking notes in class, checking reviews in the textbook, reviewing old tests, and asking the teacher.

6. **Research together.**   Arrange to have answered the questions that could not be answered by the group. The group could approach a tutor or the teacher. Or one designee from the group could do so and then inform the group.

7. **Be fair to others and to yourself.**   Be supportive of all group members, helping each get the aid he or she needs. However, do not sacrifice your own learning for that of other members.

8. **See your work.**   Work together on a whiteboard if possible. Seeing your work and moving around as you do it assists understanding.

9. **Work through issues.**   If there are problems with your group, discuss those problems with your teacher or a school counselor.

The following paragraph is a personal account of study groups from a student who matured as a person in her struggles with math. In the beginning Isabella would cry in class, feeling that she was alone in her struggle. After many semesters and repeating many math classes, she became a successful math student and loved her work.

"Sometimes when I sit down to study for a test, I really don't know where to begin at all. So this last semester I joined a study group. That really helped because I spent hours and hours with them, which was hours and hours that I wouldn't have spent at home. Plus I made an effort to prepare before we met so I'd be ready to do something when we got together. We started with Chapter 1 and studied with the instructor's worksheets. We all got copies of our tests and helped each other with the ones we missed. When I took the final, those were the things that I did the best on. When people talked about the final afterwards and what they had forgotten, I got it because we had studied it. There were nine chapters on the department final. In the beginning we did the first three chapters because they were short. Then each chapter got bigger. There were just three of us, and sometimes a fourth person who would float in and out or a different person who showed up and never came back. We reserved the room in the back of the library to use the whiteboard. We met at the table and hashed over stuff, staying an hour or two, two or three times a week. [Another] group of students met on Fridays from 9 A.M. to 12 noon. They invited me and I started going to that for an hour too. By the last week we were meeting about every day during the week."

Resource #6: Tutor

Sarah has the right attitude for getting assistance. She says:

> "Talk to everyone you can about mathematics so that you can find somebody who can explain the concept to you in a way that you can understand it. By mentioning [a] problem over and over and over again, eventually you will find somebody who can understand."

A dynamic educator and administrator with a Ph.D., Judy Schaftenaar recommends, "Talk to somebody (maybe a peer) who really is . . . obsessed with math. They might be able to talk about it in a fun way even if they are at a higher math level." Emily Meek, Ph.D., a sensitive psychotherapist and artist, has similar thoughts. She says, "Talk to people who really are excited about math. Part of getting stuck is not realizing the options you have."

Where do you find those people who are excited and obsessed with math? Where do you find a tutor? Check out the tutoring services on campus. Often there is a central tutoring center for all subjects, a tutoring center specifically for math, and subject-matter tutors in special programs. These tutors will all be free of charge to you. If you can afford a private tutor, ask every math teacher you know for a recommendation. Ask other students if they have an excellent tutor to recommend. To use the tutors on campus, go to that program, introduce yourself, and ask about their services. They will explain their hours, tutor availability, how to set up an appointment, and their expectations of you. Be responsible when you use these campus services. They do cost money even if they are free to you. When you don't show up for an appointment, someone loses money. It could be your tutor, who could have been working with someone else during the time you didn't show up for your appointment. Once you begin using the services in a tutoring center, you are not locked in. You can cancel a reasonable time in advance and you can request a different tutor.

Use this checklist to help you optimize your time with your tutor:

SHORT-TERM GOALS FOR WORKING WITH A TUTOR

1. **Find the best.**   Choose someone with whom you feel comfortable working. Find a tutor who is very knowledgeable in the area in which you need help.

2. **Use time well.**   Choose someone with whom you will focus on math—not a friend who you will be tempted to chat with rather than work with on math.

3. **Take action yourself.**   Have your tutor watch you work and correct your mistakes, not do all your work for you. Expect that you will talk more than your tutor talks. That way you will be the one learning more, not the tutor.

4. **Focus on *your* class.**   Show the tutor your class notes so that she or he can help you with the strategies your teacher uses. Be skeptical if your tutor always wants to show you a new and better way than your teacher is teaching. Your tutor should be more flexible than you. Being shown too many methods can be confusing to you.

5. **Be timely.**   Expect that both you and the tutor will arrive at the expected time.

6. **Critique the session.**   Choose another tutor if you've given your all and you find a tutor's actions unhelpful or if the tutor is unwilling to answer questions. Some tutors are trained and some are not. Some tutors are good and some are not. It is O.K. to fire your math tutor and find another.

SAC student Alex had an outstanding tutor. He describes his experience:

> "I would ask my tutor the questions that I needed to have answered and most of the time every-thing started to come together. That helped out a lot. My tutor never did the work for me. He would make me explain it to him and I think that was the key part to it. If I couldn't explain it, that meant that I didn't understand it. If I explained it, I would have confidence in myself, knowing that I could explain it to somebody else. I talked more than he did. He would just listen and correct me if I was wrong. He would say, 'That's correct' or 'You want to do this first.' If he did the work, I would always say, 'Oh, I understand it.' It would be like going to class when the instructor does the problems. I always understand them in class but when I sit down to do the homework, it's like, 'Oh, no.' Similar problems but I can't do them. There's just something that's different when somebody else does the problem for you and when you do it yourself."

Afterthought: Is There an Ideal Math Teacher/Classroom/Experience?

"A good teacher is the holiest of God's creatures. I don't think that there's another role in our culture that deserves the rank of holy except a great teacher who excites you. And the bad teachers should be made to march in chains."

Barbara Sher (McMeekin, 2000)

I love what Barbara Sher says about teachers. However, I know that whether I would be ranked holy or marched in chains would depend on which of my students you talked to. Learners and teachers are all individuals and different from each other. As a math student, remember that there are no perfect teachers and no perfect classrooms. There are, however, teachers and classrooms that can work for you.

Evaluate your math class and math teacher. If you can recognize a positive classroom environment, you may be able to help create one. It is too easy to blame the instructor for frustration with math. Total honesty and introspection pinpoint where the problem lies. The bottom line is that because no one can change others, the real power lies in changing ourselves.

It is *you* who must make your math experience worth your time and effort. If the problem is not you, you might still be able to facilitate your learning in a negative environment by observing but not absorbing and not personalizing the negativity. You cannot make your teacher change, but you can evaluate the class atmosphere.

In the next section I quote the math department philosophy from my college. As I reread this teaching philosophy, I recognize how "ideal" it is. Being human, I know that, as a math teacher, I can only aspire to these goals. I know that there are days when I am distracted or under the weather or just plain insensitive. As hard as I try to keep my classrooms always safe, encouraging, interesting, and accessible, I know this will not always happen.

As you take charge of your learning experience, you will be able to forgive your math teachers for not being perfect because you will know how to get what you need. Find a math teacher who aspires to this philosophy and the five goals at the end.

Santa Ana College Math Department Philosophy

In this math department, we believe in empowering students. Negative past experiences, learning disabilities, and current life stressors all affect a student's ability to gain access to the linear, analytic functions of the brain required to do math.

It is now widely known that Albert Einstein and Thomas Edison appeared dull and slow as students. Winston Churchill flunked English. Leonardo da Vinci, Ludwig von Beethoven, Louis Pasteur, and Hans Christian Andersen had learning disabilities. As we look out over our math students or grade their exams, we cannot know the depths of their abilities. All we know is what they can currently access.

Therefore, it is in their best interests that we provide an atmosphere that is safe and positive so that they can begin to open their minds to math. This is not to say that we "lower our standards" or that we become floor mats and "water down our courses."

It is to say that we mirror positiveness and possibilities to them. We provide them with support. We give them consistent feedback on the bits of progress that they make so that they continue to put one foot ahead of the other working their way up the math mountain.

We may be the first math teacher they ever had who believed that they could do math or the first to present it in enough different learning modes so that they could finally grasp it. We may be the first math teacher who ever gave them permission to make mistakes and to take the risks that allow them to learn.

- When we as math teachers are willing to examine the shadowy parts of our academic past and think about the courses we enjoyed the least,

- When we are willing to recognize that our math abilities gave us a certain intellectual status so that we had permission to not do so well in perhaps P.E. or English comp,

- When we are willing to admit our discipline is no better and no worse than any other academic discipline but that it currently enjoys a reputation as being the best indicator of intelligence,

then we can truly realize the incredible courage it takes for students whose skills lie elsewhere to enter our math classrooms.

Therefore it is our belief in this math department that to be truly effective with our students, we need to recognize the possibilities that are keeping our students from learning. We need to encourage, encourage, and encourage. We need to facilitate our students' use of the extra supports that we have on our campuses for tutoring, coping with math anxiety, personal counseling, and diagnosing and coping with learning disabilities. It is also helpful if we have read materials on math anxiety so that we do not perpetuate some of the negative ideas that fill students' heads and cause static, preventing clear thinking.

Being a math teacher

- who clearly verbalizes expectations and ground rules—writing them out in our course outlines,

- who gives class presentations that are well thought out and organized,

- who calls students by name and actively engages them positively in the learning experience,

- who varies classroom activities to accommodate diverse learning modes and attention spans, and

- who is knowledgeable about support services and encourages students to use them,

can go a long way toward reducing math anxiety and releasing student energy to be used on math.

We have a challenging and rewarding job to do. Isn't it wonderful?

ACT FOR SUCCESS | CHAPTER 6

1. Make a list of five specific resources (names, locations, phone numbers) where you can get answers for your questions. Refer to this list when you are frustrated. At stressful times when emotions flood your brain, written strategies can solve your dilemma.

2. Write five suggestions within your control for getting the teacher [or tutor] on your side.

3. How can you actively participate in your math classroom? Make a plan to get what you need there. Do you need to change where you sit, get to know your teacher better, reread the syllabus, spend more time preparing, list your questions before class, or set up your note taking before class?

4. Write five advantages of participating in a study group. Consider forming a study group. Make a plan or ask your instructor to help you.

MASTER MATH'S MYSTERIES

Adding and Subtracting Decimals

Recall that we have 10 digits in our number system. They are 0, 1, 2, 3, 4, 5, 6, 7, 8, and 9. In addition, our number system is a place-value system. For example, we know that the digit 1 in the number 10 is worth much less than the digit 1 in the number 10,000. The *value* of the digit depends on its *place* in the number, so we call our number system a place-value system.

When looking at a large number like 1,234,567,890, we know that the individual digits like the 7 are in specific places that mean something. Each digit has a place value that we recognize by where it is in the whole number. For example, the 7 is in the thousands place and is therefore worth seven thousand. We can see the place value of the other digits in the chart.

billions	hundred millions	ten millions	millions	hundred thousands	ten thousands	thousands	hundreds	tens	ones
1	2	3	4	5	6	7	8	9	0

When looking at a small number like 0.246815, we also have place values associated with each digit. The digit 8 in 0.246815 is in the ten-thousandths place and worth 8 ten-thousandths or $\frac{8}{10,000}$. The digit 2 is worth two ten**ths** or $\frac{2}{10}$. The digit 4 is worth four hundred**ths** or $\frac{4}{100}$. The digit 5 is worth five million**ths** or $\frac{5}{1,000,000}$. The "th" at the end of each place value name indicates that the place value is smaller than one and falls behind (to the right of) the decimal point. See the decimal-place chart.

ones	tenths	hundredths	thousandths	ten thousandths	hundred thousandths	millionths
0 ●	2	4	6	8	1	5

Note that the names for the place values as they increase to the left of the decimal point (in the first chart) are the same as the place values as they decrease to the right of the decimal point (in the previous chart), with two important differences. Those differences are:

1. The "th" endings let us know we are to the right of the decimal point, not to the left.

   Example: Compare 10 and 0.1. The "1" in 10 is "one ten." The "1" in 0.1 is "one ten**th**."

2. There is only a single ones place. When we move beyond the decimal point to the right, we start counting with the tenths place.

   Example: In the number 34.6, the "3" is "three tens," the "4" is "four ones," and the "6" is "six tenths."

When adding whole numbers like 23 + 145, we make sure to add the ones to ones and the tens to tens, by lining up the digits into columns.

Example:

```
   23
+ 145
  168
```

Notice that the 2, 4, and 6 are lined up vertically because they are all "tens." Notice that the 3, 5, and 8 are lined up vertically because they are all "ones."

Study Tip: Use graph paper for math homework and note taking. Graph paper will help you line math up vertically. Check out my favorite graph paper, called "Engineers Computation Pad."

Adding Decimals

When adding decimals, we need to make sure we are adding "like to like." This means that when we add 0.3 (3 tenths) to 0.07 (7 hundredths) we do not add the 3 and the 7 together. Adding the 3 and the 7 to get 10 would be like adding 3 dimes and 7 pennies without converting the dimes to pennies. We must add the tenths to the tenths and the hundredths to the hundredths just as we would add dimes to dimes and pennies to pennies when dealing with money. The easiest way to line like with like is to line the decimal points up vertically.

Example: Add 0.3 + 0.07

```
  0.3
+ 0.07
  0.37
```

(continued)

Example: Add 2.78 + 10.5

$$\begin{array}{r} 2.78 \\ +\ 10.5 \\ \hline 13.28 \end{array}$$

The ones line up in a nice column just like when we were adding whole numbers. (Memory tip: To make this easier to remember, we could think of the Beatles and the phrase "Line up, line up for the magical math mystery tour.")

Try these: (Hint: Line up, line up!)

1. 347 + 12	**2.** 3.47 + 0.6	**3.** 0.004 + 0.2	**4.** 14.7 + 6.201
5. 0.0034 + 4	**6.** 9.2 + 5.6	**7.** 624 + 6.24	**8.** 0.0301 + 0.0031

Subtracting Decimals

When subtracting decimals, the same ideas apply. We need to subtract like from like. First, we need to line up the decimal points so we will be subtracting tenths from tenths, hundredths from hundredths, and so on. Next, we subtract the two numbers as if they are whole numbers (remember to borrow when needed) and then make sure we carry the decimal point down into the answer.

Example: Subtract 0.3 − 0.07

$$\begin{array}{r} 0.3 \\ -\ 0.07 \\ \hline 0.23 \end{array}$$

Example: Subtract 3.47 − 0.6

$$\begin{array}{r} 3.47 \\ -\ 0.6 \\ \hline 2.87 \end{array}$$

Try these:

9. 347 − 12	**10.** 5.84 − 0.7	**11.** 0.4 − 0.002	**12.** 14.7 − 6.201
13. 4 − 0.0034	**14.** 9.2 − 5.6	**15.** 624 − 6.24	**16.** 0.0301 − 0.0031

Money and Decimals

Some people like to remember that money uses decimals. The place values for money are given in the following chart. The example in the chart is worth $87,654.32. The digit 3 represents 3 dimes and is worth 30 cents or three tenths of a dollar. The digit 5 represents 5 ten-dollar bills.

ten thousand dollars	thousand dollars	hundred dollars	ten dollars	dollars	dimes	pennies
8	7	6	5	4	3	2

Example:

Two friends were going out to dinner and decided to pool their money together. If the first person had 3 ten-dollar bills, 4 dollar bills, 2 dimes, and 3 pennies, and the second person had 1 ten-dollar bill, 8 dollar bills, and 7 dimes, how much money do they have all together?

Solution: We would pile the money together by denomination (or type) and discover that there are 4 ten-dollar bills, 12 dollar bills, 9 dimes, and 3 pennies. We could exchange 10 of the dollar bills for a ten-dollar bill. That gives us 5 ten-dollar bills and 2 dollar bills. So, adding $34.23 and $18.70 would be performed like this:

$$
\begin{array}{r}
34.23 \\
+\ 18.70 \\
\hline
52.93
\end{array}
$$

The friends have $52.93.

Student Success Story

Jazmin Hurtado

Speaking only Spanish in Mexico from age 7 to 16, math student **Jazmin Hurtado** maintained her English skills by daily watching the Disney Channel on television. At Santa Ana College, Jazmin worked her way up from prealgebra through the math sequence. Although the beginning algebra and geometry courses were actually fairly easy for her, she discovered that she had to read the book and dig in more as she progressed. Forced to reevaluate and modify her work and school schedule, Jazmin found that she needed hours of study time—often more than her study partners did. Undeterred, she depended on her own persistence and honesty about whether she had really done what was needed to learn the material. A too-full schedule hindered her from passing second-semester calculus the first time through, but she stayed in the course until the end, taking detailed notes. Those notes enabled her to pass the next semester even though she broke her foot just before the first exam.

A popular algebra assistant, Jazmin worked in the college transfer center as she completed her University of California at Irvine bachelor's degree in biology. Jazmin toyed with becoming a math instructor eventually, even though her main goal was medical school. She saw her friends taking more math and found that she was curious about what they were learning and missed the challenge. Jazmin likes math because it feels natural to her. She understands it and can picture it. She says, "Sometimes I get an answer but do not grasp the concept. That really bugs me, so I sit there and stare at the problem and do it over in my head for two or three hours until I get it. When I get it, I feel a very good satisfaction. I get that with math—not with any other subject. That's why I like math."

CHAPTER 7

Take Charge of Testing

"Whether you think you can, or think you can't, you're right."

HENRY FORD

Test time is another critical time during your semester. To conquer test taking, take charge of your preparation and performance before, during, and after your exams. When you are in the driver's seat, there is much you can do to guarantee your success. Using one or two of these test-taking strategies to take control could transform your testing experience and minimize any anxiety.

"If you give it time, you will learn math rather than not learning it and not understanding it. If you want to learn [math] the day before the test, good luck. What I've noticed is that you've got to give it time. Make time."

Enrique De Leon, math tutor

What students call test anxiety is often caused by a lack of concentrated preparation of the kind outlined in this chapter and Chapter 5. However, you can know math and sometimes still not test well. Performance anxiety can cause "static" in your brain, hindering retrieval of the knowledge you've practiced and stored. This chapter will also teach you techniques to decrease performance anxiety and to increase clear thinking. The end of this chapter is a short section of suggestions for reframing negative thoughts you may have about test taking to get a better perspective. More suggestions to take charge of lowering performance anxiety are found in Chapters 10 and 11.

The words here were chosen for the best possible perspective for test taking. Consider these words for yourself.

"Occasionally I feel a little discouraged and panicky when a test is coming up. That doesn't mean I won't do well. I need to prepare and can begin now by recognizing and doing problems that I know, and by clarifying what I didn't understand from today's class. The rest of this chapter will help me plan."

"By next week, I will understand this new material. It is new to me now. I will practice. Understanding and confidence come with practice."

"I will choose to think and act positively as I prepare for this test. I will write down examples the teacher gives us and work them along with working review problems from the ends of the chapter in the book."

"I have purposely put myself in this challenging class to grow and to enrich my life. There are many positive, concrete ways I can prepare for this test. Worrying about the test right now will not help. Worrying actually only takes energy from today. A stress-reducing walk or relaxation could help now."

"Working and practicing problems will help me know more and feel better. Carving out time periods to practice, study, and ask questions will help me feel better too, and be more prepared."

Critical Time—Prepare Before Your Exam

Test preparation cannot begin the day before the test. Active preparation begins ahead of time when you choose the best actions for learning math and use your resources. As you read these strategies for taking charge of test taking ahead of time, check those that you are willing to consider before your next exam.

1. Work Problems

Diligently prepare and practice. The rest of this chapter cannot help during the exam if you haven't prepared your brain by learning the concepts and procedures for the test. This book is filled with ways to learn math with confidence. Chapters 4, 5, and 6 taught you how to study, read your textbook, take notes, make note cards, and get your questions answered.

Go over the chapters in your math book and class notes that will be covered. Look for the key ideas and skills. Pull out any note cards you have made or make note cards now with vocabulary, concepts, or problems. Look over what the instructor has told you will be on the

SUCCESS STRATEGIES BEFORE THE EXAM

1. Work problems.
2. Predict the test problems.
3. Make and use note cards.
4. Do a dress rehearsal.
5. Anchor success to your day.
6. Prepare your body along with your mind.
7. Learn the stress-reducing Relaxation Response.

exam. Be sure that you can do each type of problem and be sure that you will recognize the vocabulary of the directions so you will know when to do each procedure.

Prepare one special index card with formulas, facts, or examples you would wish to review immediately before your test. Use this card to review and "dump data" on your test form.

Set aside time during the week before the exam to work the chapter test or review problems at the end of each chapter in the text. Work the examples that you copied from your instructor in class.

If you find yourself procrastinating, go to a study group or a tutor. Or make an appointment to go to the library or math study center for a specific period of time. Show up and start working. You don't have to "feel" like working; just do it and the feelings may follow—or not, but either way you will learn. Set short-term, immediate goals for each study session so that you are not overwhelmed and so that you feel accomplishment afterward. Take short breaks to refresh as you study.

2. Predict the Test Problems

During the first week of class, set aside a special page or section in your notebook as a "Test Prediction Page." On it, write:

- Problems or concepts your teacher says are important or will be on exams.
- Problems or concepts you yourself believe will be on exams.
- Problems or concepts your tutors or classmates predict will be on exams.

Continually question others and the material itself as you work it to understand what is important to know and remember. Guard this Test Prediction Page and use it for test review.

3. Make and Use Note Cards

Prepare Question and Answer note cards using problems from each section of your book. You can choose these problems from your Test Prediction Page just described, from your copied class notes, or from the review sections at the end of the chapter. Put the problem on one side and show the work on the other side of the card. In addition to problem cards, make vocabulary cards with the word on one side and the definition on the other side. Prepare informational note cards listing important concepts and procedures from the material that will be covered on your exam. Carry your note cards with you. Test yourself once or twice a day by looking at the question. Answer without looking at the back, perhaps by writing out the solutions or definitions or steps. Always get feedback by checking your responses with the back side of the card. Keep these note cards throughout the course to help you prepare for the final exam. (See Chapter 5 for help in making note cards.)

4. Do a Dress Rehearsal

Write your own pretest. Use your Test Prediction Page along with your class notes and textbook to choose two or more varied problems from each section—especially the later sections in each chapter. Choose problems for which you have access to the answers. As you copy each problem, note the page and problem number and make a separate answer key. At least two days early, take your practice test as a dress rehearsal in a quiet setting. Do not use your book or your notes. When you finish, correct your test and carefully rework the problems you missed. If you still don't understand concepts or procedures, show your test to your instructor or tutor and ask for help. During the real test, you will likely be pleasantly surprised to find how closely you have guessed the content of the problems.

5. Anchor Success to Your Day

Plan ahead of time what to take, what to wear, when to arrive, and where to sit on the day of the test to "anchor" you to yourself at your best. Be creative in providing yourself with the best and safest possible environment so you can concentrate your energies on the test itself.

- Prepare a test kit of "tools" to take with you to every exam. Include sharp pencils (especially your favorite ones), pens, erasers, correction fluid, a ruler, tissues, a bottle of water, a calculator, spare batteries, and so on.

- Make sure you know where and when your exam will be and how you will get there. Come early to the exam to pick the best spot for yourself. Make sure that you have space around your desk so you can move if you need to. If other students distract you, sit in the front corner farthest from the door.

- Take your favorite pencil or "lucky rock." Boost your confidence by pocketing a special note or card from a friend reminding you of your value to others and your life beyond this exam.

- Wear clothing in which you feel good. You might wear comfortable, nonbinding clothing, or a sharp professional-looking outfit, or your "lucky shirt." Take a jacket or sweater to keep you warm in an air-conditioned classroom so oxygen-rich blood flows to your brain during the whole exam period.

If something happens and you find yourself rushing to your test, you may be interested to know that some studies have shown that a brisk walk before an exam has boosted test scores, maybe because the increased blood circulation accelerates oxygen flow to the brain. Whatever happens on your test day, tell yourself that these conditions only help you concentrate more and think your best about math.

6. Prepare Your Body Along with Your Mind

Scientists have biological evidence that your mind is intimately intertwined with your entire body.

- Consider the foods that work best for you. Sugars and refined foods are known for reducing mental clarity. Notice whether caffeine helps or hinders; it works for some and not for others. Protein appears to be essential for clear thinking. Carbohydrates may make you tired. Observe the foods that keep your body calm and alert and then control your food intake on test day. Pack your own food to take with you so you are not dependent on vending machines or fast-food restaurants.

- Get the optimal amount of rest during the week before the test. This is individual and personal, but it is important to have the amount of sleep that keeps you sharp.

- Continue daily walks or bike rides or whatever aerobic exercises you use regularly. Do not neglect these stress-reducing/mind-relieving activities as you concentrate on math. If you do not have any regular exercises, choose some mild form of body movement such as stretching or walking around in your home or yard.

7. Learn the Stress-Reducing Relaxation Response

The Relaxation Response is the antidote to experiencing "fight-or-flight" during a test. The fight-or-flight instinct automatically releases adrenaline into your body, increasing your heart rate, respiration, and metabolism to produce the energy you need to cope with emergencies. This

THE RELAXATION RESPONSE

1. Sit quietly and comfortably with your eyes closed.

2. Systematically relax your muscles, beginning with your feet and moving slowly up your body. Tense each muscle group (feet, calves, thighs, buttocks, stomach, chest, shoulders, arms, hands, shoulders again, and face) one group at a time, and then give those muscles permission to release and "let go" as much as possible. Allow "streams of relaxation" to flow throughout your body.

3. Breathe slowly in through your nose, counting to 5, and then breathe out through your mouth, counting to 10. If you feel dizzy, stop breathing for a moment or breathe into a paper bag.

4. As distractions come into your mind, let them go. You can take care of them later. For now, just "be" in the moment as much as possible. Simply relax as much as you can. You might imagine yourself in a safe, comfortable, and caring place just noticing what you hear, see, and feel while you are there. Any type of slow stretching, yoga, or meditation as you breathe is helpful.

5. Continue to relax and breathe for 10 to 20 minutes once or twice a day. (Do not practice these deep-breathing exercises while you drive.)

automatic response saves people in accidents, bear attacks, avalanches, and fires. On the down side, during an exam, the fight-or-flight response is a nuisance, preventing calm, clear thinking.

Because most math exams do not involve true emergencies such as accidents, bears, avalanches, or fires, having excessive adrenaline in your body simply makes sitting attentively and concentrating on the challenges of the test difficult.

The good news is that the antidote to the fight-or-flight syndrome was researched and documented by Dr. Herbert Benson of Harvard in 1975. He calls it the Relaxation Response. Although developed decades ago, it is still cited as a very effective stress reducer. The more the Relaxation Response is practiced, the more automatic and natural it becomes. The Relaxation Response, when practiced daily, decreases adrenaline and releases other body chemicals that promote relaxation, focus, and clear thinking. Professional athletes, musicians, and highly productive people use this technique often.

Once you learn and daily practice the Relaxation Response, your body relaxes easily as you simply take a deep breath at key moments during your test, including as you begin, when you get stuck, or when you need a break. Using relaxation consciously during an exam frees the thinking part of your brain.

Critical Time—Focus During Your Exam

You can do much to set the tone for the day and the time of your exam. It is a special day. Excitement is natural. It is a day to speak kindly and encouragingly to yourself. It is a day to walk past any fears and to be very objective that you will do the best that you can under the circumstances. Here are strategies to make the exam a successful, even *enjoyable experience*.

SUCCESS STRATEGIES AT YOUR EXAM

1. Ignore others before the test.
2. Do a data dump.
3. Scan twice.
4. Strategize.
5. Use time wisely.
6. Trust your subconscious mind.
7. Ignore others during the test.
8. Roll over distractions.
9. Take minibreaks.
10. Ask questions.

1. Ignore Others Before the Test

Do not absorb other people's anxiety like a sponge. Some people deal with their anxiety by verbally expressing their worst fears and concerns in an excited or hysterical way. If these conversations bother you, protect yourself by building an invisible shield around yourself. Merely observe their anxious state or walk away.

It is perfectly O.K. to avoid talking with anyone and to sit by yourself before the exam so that you can breathe deeply, look over your review note card, and focus. Or you may find that telling a few jokes or laughing with others relaxes you. This is a time for you to care for yourself in whatever way you need. Choose what you do and with whom you speak before each exam. You will feel more in charge.

2. Do a Data Dump

Bring an index card of formulas or facts you find difficult to remember. Look at them before the test. Visualize the formulas and facts going into a holding tank in your brain. *If you are not allowed to use notes on the exam, be sure to put the list far away so that your honesty is not questioned.* When you receive your test, quickly write these formulas or facts on your exam paper. Now you do not have to expend any energy trying to recall them later when you need them.

3. Scan Twice

Scan the entire test at least twice—once at the beginning and once at the end. At the beginning, notice the kinds of problems, how many problems, and with which problems you would be comfortable starting. Write down any facts or formulas on the exam that are brought to your mind with this overview. At the end of the test, scan again to **be certain you have worked every problem**. Write something for each problem even if you only recopy it and write down any intuitive ideas you have about it. Instructors often give points for "trying." It might even happen that once you begin, you are able to work the problem.

You might develop your own test rituals or procedures for the beginnings and endings of tests. Then you will have a routine going and won't become rattled. For example, at the beginning, your test ritual could be to do a data dump, write your name on the test, take a deep breath, put your feet flat on the floor, scan the test, jot down any more facts or formulas that come to mind, take another deep breath, and begin the problem you are most comfortable doing. Perhaps at the end, your closing test ritual could be to take a deep breath and a minibreak, scan the test, work any skipped problems, rework any problems with question marks, check all problems by reworking them, use all of the test time, and choose to be the last person to hand in your test.

4. Strategize

You do not have to take the exam in the order it appears. Do the problems and questions that you like first. Make tiny pencil marks by those questions to which you want to return. Be sure that you try all of the problems even if you can only set them up. Turn in any of your scratch paper. Your teacher may give you partial credit for setting up problems.

Don't watch the clock more often than necessary. The instructor will stop you when the exam is over.

If some of the questions are multiple choice, eliminate the obvious wrong answers first and then do the work until you can choose the exact answer. Be sure to simplify your answers. Be certain to answer all questions asked in each problem. Ask yourself if your answers are reasonable.

5. Use Time Wisely

Don't work on one problem for a long time. Often a question further into the exam will act as a "key" to unlock a previous problem. Tell yourself that you have all of the time you need. Let go of the rest of your life during the exam. You can deal with all that later. Be sure to save time to check over your problems at the end. Also, glancing over your completed test may help you spot a problem that you accidentally skipped.

6. Trust Your Subconscious Mind

Let each question reach into your mind for the answer. Remind yourself that you know everything you need to know for now. If a problem makes no sense, read it and go on. **Ideas will come to you** as the problem sinks into your subconscious mind and you continue with the test. This is called parallel processing, and you do it all the time. You may wish to just read over any essay questions or word problems early in the exam to let your subconscious mind begin to understand and process them. You will collect ideas as you work the other problems and you will be more prepared when you are ready to answer them.

7. Ignore Others During the Test

You do not need to compete with anyone except yourself. No matter how many pages anyone else turns or how much noise they make or how soon they leave the exam, you will not know how well they performed. Don't compare. You need this energy to do your work.

Ignore people who leave. Often one of the first three persons to leave an exam gets a very low score. **Often one of the last three persons to leave gets a very high score.** Take your time. Let other people's behavior go for now. I suggest that you actively choose to be the last student leaving the test. Tell your friends not to wait and that you will be staying until the end. That choice will prevent the actions of others from distracting or pressuring you.

Earplugs shut out most excess noise. If you use them, notify the instructor so that you will not miss any announcements. Sit where you can't see others. If possible, keep your desk clear of touching other desks so you are not disturbed. If you notice a distraction, allow yourself to notice it and then let it go. Allow the instructor to take care of problems in the room. You need only take charge of yourself and your performance right now.

8. Roll Over Distractions

When you feel stuck or tense or when you are disrupted, take a deep breath in through your nose and out through your mouth and place your feet flat on the floor. Let "everything" go as you expel the air, and then place your focus back on your work. (The more you have practiced the Relaxation Response before the test, the more you will be able to relax during the test.)

9. Take Minibreaks

Take 20-second time-outs during the exam to close your eyes, sit up tall, breathe deeply, or stretch your neck and arms. Tell yourself that you have all the time that you will need. A few isometric exercises can release tension too. With your feet flat on the floor, take hold of the sides of your chair and pull up. You might massage your temples, scalp, and the back of your neck to increase blood flow carrying oxygen to your brain to help you think more clearly. Close and relax your eyes by covering them for a short time with your cupped palms. Draw a small empty box on your paper. Imagine it is a TV or computer screen and picture a wonderful, relaxing scene.

10. Ask Questions

Ask the instructor questions as needed. Raise your hand and keep working. Let the instructor come to you. The worst thing that can happen is that she will say, "I can't answer that question." Often just giving yourself permission to ask a question allows you to think differently and to figure it out on your own.

Critical Time—Refocus After the Exam

 SUCCESS STRATEGIES AFTER THE EXAM

1. Let the results go for now.
2. Stay calm with results.
3. Analyze errors for feedback.
4. Evaluate your study behaviors.

1. Let the Results Go for Now

Test taking uses a lot of energy. Immediately after the test, you may feel low and off balance. You may wish to pass up discussing the exam with others so you can take care of yourself. Going to the bathroom, drinking water, and eating something can help you feel normal again. You may have put much of your life on hold to prepare for this exam. Refresh yourself and get your life back. You can deal with the test results later when your priorities are in order again.

2. Stay Calm with Results

Do not compare results with anyone else. These results are between you and the instructor. A good grade does not mean you are a wonderful, great person any more than a poor grade means that you are a horrible person. The test was just a onetime evaluation of how you were able to do certain problems with the understanding you had at the time. If need be, you can learn the material and repeat the exam or the course. Test results are not necessarily fun, but they are not the end of the world *and* they mean nothing about your worth as a person.

3. Analyze Errors for Feedback

Treasure your returned exam. It is rich with feedback on what you know now and don't know now. It can also be used to predict the format of future tests with the same instructor. When you first look at your exam, expect some "emotional flooding." You may feel very critical of yourself or discouraged. That will pass if you *keep looking*. Review your exam and learn the material that you missed. This is a great opportunity for you to get feedback about your learning.

Notice the kinds of errors that you made:

- Which errors were caused by your not knowing the material from lack of studying?
- Which errors were caused by your misunderstanding a concept or procedure?
- Which errors were caused by your not remembering something that you learned?
- Which errors were caused by your misreading the problem?
- Which errors were caused by your not reading the directions?

Now work all of the missed problems correctly. Staple that work to your test for use when preparing for your final exam or even when reviewing before your next math course.

Knowing the types of errors that you make gives you direction for preparation next time. Because new math material depends upon what you've covered previously, spending quality time with old exams will pay you big dividends. Keep your old exams and corrections for practice and review. They will be especially useful two weeks before your final exam. Imagine how confident you'd feel if you studied the errors and retook all of your chapter exams a few days before your final exam. Do it!

4. Evaluate Your Study Behaviors

Completing this evaluation will help you objectively evaluate your exam experience and prepare for the next. Be honest—these answers will give you a reality check about what you can do differently to obtain higher scores. Remember that it is not how smart you are; it is whether or not you are persistent and do what needs to be done to learn the math.

Exam Evaluation

❐ Are you satisfied with this test score? If the answer to this question is "no," continue to answer the other questions in this list to point to possible changes you can make before your next exam.

❐ Did you do the assigned homework for the test material?

❐ Did you attend every class session before the test?

❐ Were you on time to class and prepared with your paper, pencil, and textbook when class began?

❐ Did you take thorough class notes, copying down what the instructor wrote and said, including all the examples?

❐ Did you complete your homework as soon as possible after class?

❐ Did you write a practice dress rehearsal test, take it, and correct it before the exam?

❐ Did you ask questions on homework problems or concepts that you did not understand?

❐ Did you have a regular time and place to do your math studying?

❐ Did you use the tutoring services on campus?

❐ Did you actually study for the exam by working problems from the book and your notes?

❐ Did you practice the test-taking strategies suggested in this chapter?

❐ Did you consult your instructor, tutor, or fellow math students when you needed outside input or assistance?

❐ Did you take care of your body by eating nutritiously and getting sufficient rest during the week before and the day of the test?

❐ Did you practice the Relaxation Response daily in preparation for the exam?

❐ Did you consciously take deep breaths and relax during the exam?

❐ Did you choose your classroom seat to avoid distractions?

What actions will you take this week to ensure a higher score next time? Use your "no" answers to set achievable short-term goals for bringing flow to math and improving your math work. These goals could be as follows:

SHORT-TERM GOALS

1. I will make time immediately after class to start my homework.
2. I will make my notes more complete in class and rewrite the examples after class.
3. I will be prepared each day with pencil, paper, and book when class begins.
4. I will ask homework questions in class.

Reframe Any Negative Test Thoughts

The reframe of a thought is another thought that is equally true and possible in the situation. However, an effective reframe of a negative thought opens up the possibilities and provides a larger perspective. Read these Test Thought Reframes out loud often. (Do not read the Negative Test Thoughts aloud!) More assistance reframing overwhelming negative thoughts is available in Chapters 10 and 11.

Negative Test Thoughts	Test Thought Reframes
I will fail my exam.	I cannot predict my test score because I am unable to see into the future. I can work now getting help when I need it and learn more so that I am prepared for testing.
I will panic during the test.	I cannot foresee my reactions during the test, but I do know that emotions will come and go during the exam time. If panic occurs, it will pass through and I will continue to focus on the work itself.
I will forget everything.	With my work before the exam, I will make solid brain connections that I can access during the testing period. The exam itself will be filled with cues—words and symbols—that will trigger formulas and ideas that I have practiced.
Math tests are tricky.	My preparation will be thorough. I will understand basic processes and I will practice to gain confidence to make any "tricks" disappear.
There is so much I don't know.	I work to increase my knowledge in math day by day. It will always be the case that there is much anyone doesn't know. I don't have to know everything; I am learning what the test covers.

ACT FOR SUCCESS | CHAPTER 7

1. Make 20 note cards with problems that span your class notes as described in point 3 of "Success Strategies Before the Exam." Work through your note cards twice a day, quizzing and checking yourself to review for your exam.

2. Write, take, and check a dress rehearsal test as described in point 4 of the "Success Strategies Before the Exam" box. After your exam, highlight all of the problems on your dress rehearsal test that were similar in concept to those problems on the real exam.

3. Write down five specific actions that you will do at your next exam to foster success. Plan carefully how to execute these strategies. How will you remember what to do?

4. Answer the "after the exam" evaluation. Write five short-term goals you can set to raise your test results.

MASTER MATH'S MYSTERIES

Symbols, Signs, and Signed Numbers

Important Symbols in Mathematics

Sometimes the individual symbols used in mathematics have several different meanings. We see similar instances in English. Read the following sentence out loud:

"The soldier wound the bandage around the wound."

Notice that two of the words have identical spelling and yet you pronounced the two words differently. These words look the same but are pronounced differently and mean different things. You recognize the difference by the context of the sentence. Mathematics has similar issues.

Many people complain of the multitude of symbols used in mathematics. The "−" symbol is confusing for many people. This section will discuss "−" from many perspectives.

Consider this problem:

$$-(-2 - 4)$$

This math expression has three symbols that look identical, but each has a different meaning. The context of the symbol within the expression lets you know what it means. This math expression is read "Opposite of the quantity of negative two subtract four." What does this really mean? To understand, we discuss signed numbers.

Signed Numbers

All numbers are signed numbers. The sign, positive or negative, represents direction. A positive direction is opposite a negative direction. For example, 2 could represent traveling 2 miles east on a subway train and −2 could represent traveling 2 miles west in the opposite direction. Or 2 could represent winning 2 dollars and −2 could represent losing 2 dollars. Or 2 could represent moving up 2 floors on the elevator and −2 could represent moving down 2 floors.

Historically the Chinese used rod numerals, with different-colored rods representing positives and negatives. In this discussion we use colored beans to represent the positiveness and negativeness of the numbers. We use light-colored beans to represent positive numbers and dark-colored beans to represent negative numbers. Our signed numbers now have two attributes: quantity (How many?) and color (What sign?).

We let dark beans represent negative numbers.   = −3

We let light beans represent positive numbers.   = 3

Adding Signed Numbers

Adding Positive Numbers (or Light Beans)

Example:

3 light beans   +   5 light beans   =   8 light beans

3   +   5   =   8

Try these. Answers to Master Math's Mysteries are found in the Appendix.

1.

_____   +   _____   =   _____

Draw beans for numbers 2–4:

2. 5 + 1 = _____   **3.** 3 + 6 = _____   **4.** 4 + 3 = _____

Adding Negative Numbers (or Dark Beans)

Example:

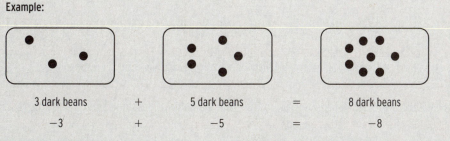

3 dark beans   +   5 dark beans   =   8 dark beans

−3   +   −5   =   −8

Try these:

5.

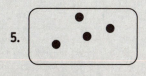   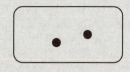   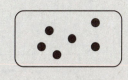

_____   +   _____   =   _____

Try these by drawing beans:

6. −6 + (−2) = _____   **7.** −3 + (−2) = _____   **8.** −3 + (−4) = _____

(continued)

Zero Pairs.   For both dark and light beans together, we use what we call "zero pairs." If I give you a dollar (you gain $1) and then take it back (you lose $1), you started with zero and ended with zero.

= 0

The helium atom has 2 electrons (negative charge) and 2 protons (positive charge), which balance out the charge of the atom to zero.

= 0

If you dig out 5 buckets full of dirt and put 5 buckets full of dirt back in the hole, the ground should look the same.

= 0

Let us agree that one pair of a dark bean and a light bean combine to form zero. We call these combinations "zero pairs."

Example: Combine zero pairs before you record the result:

= 2

Notice that we have 3 light beans and 1 dark bean. Together one light bean and the dark bean make a zero pair. We cross the zero pair out and find that we have 2 light beans remaining.

In each picture, cross out the zero pairs and write the name of the signed number for the remaining unmatched beans below the box.

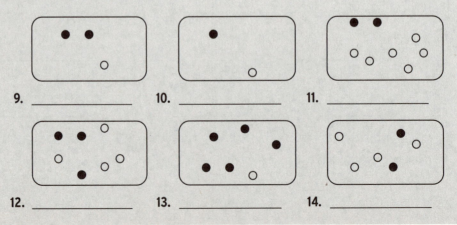

9. _____   10. _____   11. _____

12. _____   13. _____   14. _____

Adding Positive and Negative Numbers (or Light and Dark Beans)

Example 1:

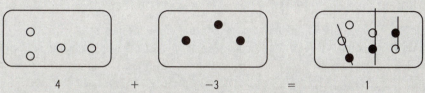

$$4 \quad + \quad -3 \quad = \quad 1$$

We add 4 light-colored beans (positive 4) to 3 dark-colored beans (negative 3) and discover 3 zero pairs. Recall that a zero pair is one light bean and one dark bean. When we have lined out the zero pairs, our answer is 1 light bean, or 1.

Example 2:

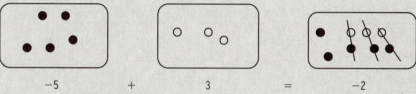

$$-5 \quad + \quad 3 \quad = \quad -2$$

We add 5 dark beans (negative 5) to 3 light beans (positive 3) and discover 3 zero pairs. We line out the zero pairs to find our answer is 2 dark beans or −2.

Try the following by lining out the zero pairs in the solution box to the right:

15.

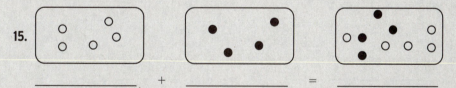

_____ + _____ = _____

Draw beans in the boxes to do problem 16. Line out the zero pairs in the box on the right.

16.

$$-6 \quad + \quad 3 \quad = \quad \text{_____}$$

Try numbers 17-25 by drawing beans and marking out the zero pairs:

17. $3 + (-7) = $ _____   **18.** $-4 + 6 = $ _____

19. $9 + (-2) = $ _____   **20.** $8 + (-5) = $ _____

21. $-7 + 2 = $ _____   **22.** $-8 + 4 = $ _____

23. $9 + (-3) = $ _____   **24.** $4 + (-6) = $ _____

25. $-2 + 6 = $ _____

(continued)

Important Note: For problems 17-25, circle the number that has more beans of a single color. Notice that your answer has the same sign as the number you circled. When you are adding, your answer will always have the sign of the number with the larger quantity.

Subtracting Signed Numbers

Subtracting Positive and Negative Numbers (or Light and Dark Beans)

To subtract, we use the "take away" method and again we use light and dark beans. To show the subtraction, we draw a circle around the number that is taken away.

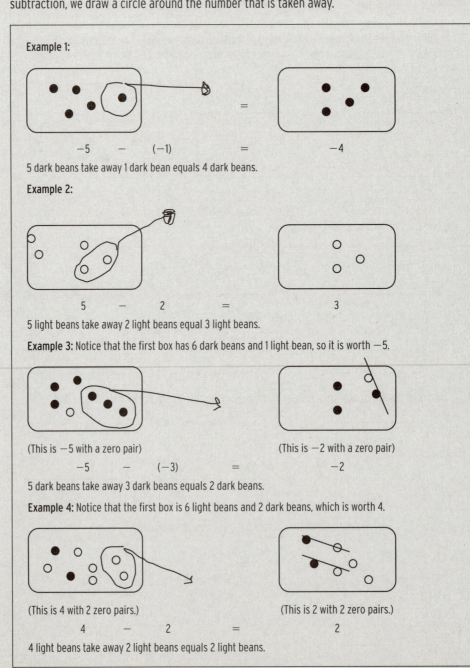

Example 1:

$-5 \quad - \quad (-1) \quad = \quad -4$

5 dark beans take away 1 dark bean equals 4 dark beans.

Example 2:

$5 \quad - \quad 2 \quad = \quad 3$

5 light beans take away 2 light beans equal 3 light beans.

Example 3: Notice that the first box has 6 dark beans and 1 light bean, so it is worth −5.

(This is −5 with a zero pair)   (This is −2 with a zero pair)

$-5 \quad - \quad (-3) \quad = \quad -2$

5 dark beans take away 3 dark beans equals 2 dark beans.

Example 4: Notice that the first box is 6 light beans and 2 dark beans, which is worth 4.

(This is 4 with 2 zero pairs.)   (This is 2 with 2 zero pairs.)

$4 \quad - \quad 2 \quad = \quad 2$

4 light beans take away 2 light beans equals 2 light beans.

In numbers 26-28, write the problem shown in beans as a problem in numbers:

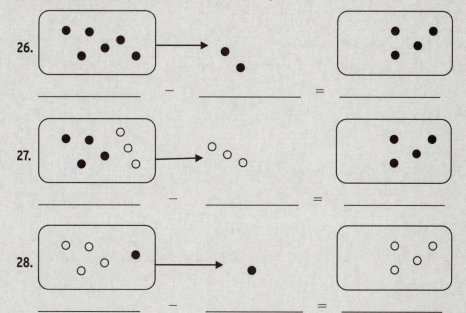

26. _____ − _____ = _____

27. _____ − _____ = _____

28. _____ − _____ = _____

Important Example:
How would you do the following: $4 - (+6) = ?$
This problem is 4 light beans minus 6 light beans. We begin with 4 light beans.

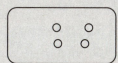

Can you take away 6 light beans from 4 light beans? Not yet! We need more light beans. **To get more light beans, add zero pairs until we have 6 light beans.**

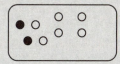

We make the +4 look different by adding 2 zero pairs of light and dark beans while keeping the same value +4.
Notice that you can now subtract 6 light beans. _____

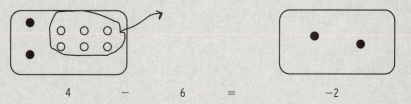

    4   −   6   =   −2

(continued)

Add the needed zero pairs to each set in numbers 29-30 and show how the subtraction problems can be done:

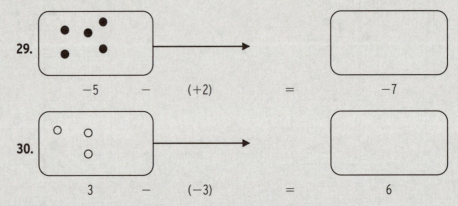

29.

-5 $-$ $(+2)$ $=$ -7

30.

3 $-$ (-3) $=$ 6

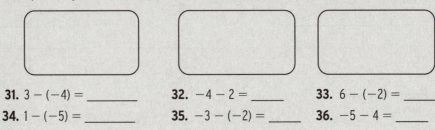

(Note that the answer to number 30 is not zero.)

Try these by drawing beans like the problems above. (Hint: You may need to add zero pairs.)

31. $3 - (-4) =$ _____   **32.** $-4 - 2 =$ _____   **33.** $6 - (-2) =$ _____

34. $1 - (-5) =$ _____   **35.** $-3 - (-2) =$ _____   **36.** $-5 - 4 =$ _____

Back to the Beginning

Remember our original example with the three uses of the "−" symbol?

That problem was: $-(-2 - 4)$

We can now explain what the three different "−" symbols in this example mean.

The Meaning of the Symbol "−"

In the order the "−" symbols appear in the problem $-(-2 - 4)$,

1. A "−" symbol by itself means **opposite**, which means the same quantity but in the opposite color. For example, the opposite of 7 or 7 light beans is 7 dark beans, or -7. The opposite of 3 dark beans or -3 is 3 light beans, or 3.

2. A "−" symbol with a number after it, but *without* another number immediately in front of it, means **negative**. A negative number represents the indicated quantity of dark beans. For example, -5 is five beans that are colored dark.

3. A "−" symbol between two numbers means **subtraction**. You subtract or take away the second amount from the first amount. For example, $7 - 5 = 2$

To Simplify — (−2 − 4)

Following the order of operations, we do the inside of the parentheses first.
So, −2 take away a positive 4 becomes a −6.

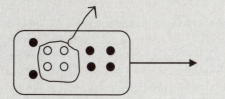

(Notice that we had to add 4 zero pairs to get 4 light beans to take away positive 4.)
Finally, the opposite of −6 is 6.
Or − (−2 − 4) = − (−6) = 6

Here are some good practice problems. Try these while thinking beans!

37. $-2 - (-5) =$ **38.** $3 - (-6) =$ **39.** $-4 + (-5) =$

40. $5 + (-7) =$ **41.** $-6 + (-2) =$ **42.** $2 - 8 =$

43. $-7 - (-2) =$ **44.** $-4 - (-9) =$

Look Back—Comprehensive Review for Your Practice

Simplify:

1. $5 + (-9)$ **2.** $110 + 1,890$

Fill in the blank:

3. The sum of 9 and −4 is _____.

4. The difference of −2 and 6 is _____.

Perform the indicated operation:

5. $\dfrac{1}{12} + \dfrac{1}{2}$ **6.** $73 + 29$

7. $24 - 2(14 \div 2) + 5$ **8.** $4 - \dfrac{1}{2}$

Perform the indicated operation:

9. $5.1 - 4.2$ **10.** $53 \cdot 10$

Student Success Story

Carlos Ordiano

An outstanding musician, pianist, and composer in various genres, Carlos Ordiano excelled in his music classes but put his general education courses off to travel and perform. With renewed drive, Carlos returned to college with the goal of completing his degree in music. It was in his intermediate algebra class that Carlos learned about Mind Mapping. His class notes showed Carlos's methodical incorporation of Mind Mapping techniques. First, he added color. Next, he added little doodles that expanded the meaning of the course content. Eventually he added arrows and connectors that showed the relationships of the concepts putting together important math topics. Studying became a joy because Carlos learned as he drew. A hard worker, Carlos shared his knowledge and his memory techniques with fellow students, even writing a few songs to help the class remember formulas. After passing intermediate algebra, Carlos took statistics and thrived as he turned all of the statistics concepts into symbols and color synthesizing their relationships with his maps. Then he turned his new note-taking skills to his other coursework, especially the concepts that he found difficult to understand and to remember. Now, with a full scholarship in the jazz studies program at California State University at Long Beach, Carlos has three sketchbooks filled with hundreds of detailed and useful Mind Maps that helped him learn all of his general education coursework (including statistics, political science, critical thinking, world music, speech, astronomy, and debate). Carlos's advice to struggling math students is to take the time to make the coursework meaningful to themselves and to use the math study center. In Chapter 8, see how Carlos has Mind Mapped the memory techniques.

CHAPTER 8

Make Strong Math Memories

"Every individual is a marvel of unknown and unrealized possibilities."

JOHANN WOLFGANG VON GOETHE

Remembering is not about memorizing. It is about making strong, long-lasting, and accessible brain connections—laying down enduring pathways in your mind. This chapter contains many hints on how to be a better math student by strengthening your math memory.

How Your Brain Remembers

In simple terms, your brain is made of billions of **neurons** (brain cells) that store your knowledge. These neurons are connected to one another by countless **dendrites**. What you "know" or "think" is made of a network of neurons and dendrites that is traveled electrically and chemically across the **synapses**, where the dendrites connect to one another (Figure 8.1). When you learn something, your brain grows new dendrites between the relevant neurons. In the beginning, those dendrites are like the roots of a young plant. They are weak and tenuous. With continued use, those dendrites grow stronger. Eventually they **myelinate** with **myelin** (a protective coating that fosters faster and more accurate thinking) if you think the same thoughts or do the same thing many times.

When you want to remember something, choose activities that will activate the network of neurons and dendrites in your brain. The more actions you take to travel the network, the more established it will become *and* the easier it will be for you to

Figure 8.1   *Neurons, Dendrites, and Synapses in the Brain*

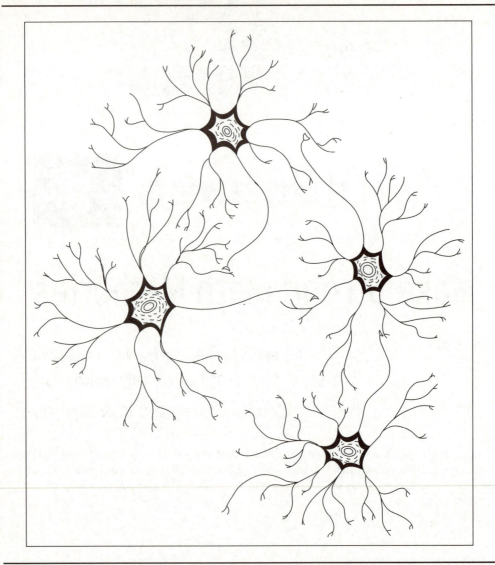

remember. This chapter describes actions that you can take to consciously improve your recall of what you need for math.

Math students often believe they must memorize everything. This is not true. Learning math is about finding meaning and fostering those dendrite networks in your brain. This involves understanding and then practicing in as many ways as possible. It is important for you to know that dendrites are specific to what you were doing at the time they grew. If you were working a math problem, the dendrites help you work a math problem. If you were just sitting and listening in math class, the dendrites formed help you sit and listen to math problems.

You want your brain connections to last long term. They store your math vocabulary, concepts, and procedures. It is those long-term connections that will assist you in making future mathematical insights and solving problems.

Your capacity for long-term brain connections is practically unlimited. This means that you have plenty of space to store mathematical ideas. However, those pathways rearrange themselves, update with new information, and deteriorate constantly. This means that your math memories get changed and confused unless they are *made carefully and reinforced by constant practice.*

This chapter follows a three-step process for forming long-term connections and gives you activities for each step of the way so that your long-term math memories will be lasting and accessible. The last section will teach you how to Mind Map—a multisensory technique that can help you synthesize your ideas in a memorable way (see Figure 8.2).

Three steps you must follow to remember math are:

- First Step—Intake: Take math into your mind memorably.

- Second Step—Storage: Store math solidly and permanently in your mind.

- Third Step—Recall: Retrieve the math you already know efficiently and reliably.

Figure 8.2   *Mind Map of Three Steps for Remembering Math*

First Step—Intake: Take Math into Your Mind Memorably

Make strong brain connections to start! Here are eight techniques for doing just that. These activities will help you form solid long-term math pathways in your mind by helping you find meaning and connections with the material. Read about each suggestion and check the ones you plan to use soon (see Figure 8.3).

 HOW TO TAKE MATH INTO YOUR MIND MEMORABLY

1. Pick and choose.
2. Get feedback soon.
3. Write or draw what you see and think.
4. Observe details.
5. Be active with what you want to learn.
6. Chunk the information.
7. Tune everything else out.
8. Have fun.

1. Pick and Choose

Choose to spend most of your time on what is most important, as pointed out by your teacher. Use your class notes and chapter review problems. Highlight important facts. Make a record (notes, note cards, or a Mind Map) of what you have assessed as important to remember. To review what is important, summarize your choices on note cards, as described in Chapter 5.

2. Get Feedback Soon

How do you know you really understand? Use what you learn immediately to clarify that it is correct. You may have misunderstood and formed incorrect connections in your mind. The longer you practice incorrectly and travel the wrong brain pathway, the more difficult a mistake is to correct. Chapter 2 has an entire list of feedback activities to find mistakes and correct them. Here are examples of how to get feedback:

- Work three or four homework problems, and then compare your answers with those in the back of the book.
- Show your work to other students.
- Teach what you learned to another math student in your class.
- Check to see if your answer works in the original problem.

3. Write or Draw What You See and Think

Visualize it and say it in your own words. The activity of writing or drawing helps you find meaning and connections with what you want to learn. It will help you notice detail. Look for unusual, vivid relationships between nonmathematical ideas and mathematical ideas. For example, the number pi, or π, is used with circles to find the circumference (the distance around the outside of the circle) and the area (the surface inside the circle). To remember pi, think of the fact that circles are round like pies.

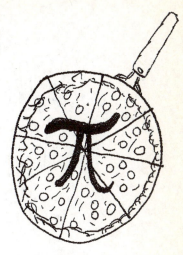

4. Observe Details

How can you be observant? Read on for hints for observing sensory data, key words, patterns, similarities/differences, and the context.

Observe key words.   If you notice key words and learn their meaning, you will organize your overall knowledge because new learning will connect with those key words. The words *sum, difference, product,* and *quotient,* described in "Master Math's Mysteries" in Chapter 3, are key words that tell you which operations to use—add, subtract, multiply, or divide.

Observe patterns.   Events that repeat themselves occur often in math. Look for those repetitions to understand what is happening overall. Story #9 in Chapter 9 illustrates the use of solving a problem by observing patterns. Here are two more examples of math patterns:

- Any number that divides evenly (with no remainder) by the number 5 ends with a 5 or a 0. The numbers 510, 275, and 1,365 can be divided evenly by 5. The numbers 712, 1,478, and 653 cannot because they do not end with a 5 or a 0.
- Numbers are divisible by 3 if the sum of their digits is divisible by 3. For example, 276 is divisible by 3 because $2 + 7 + 6 = 15$ is divisible by 3. Another example: 844 is not divisible by 3 because $8 + 4 + 4 = 16$ is not divisible by 3.

Observe similarities and differences.   To learn something, relate it to things you already know, and look for ways it is the same or different.

- -8^2 and $(-8)^2$ look very similar, but -8^2 is -64 and the other, $(-8)^2$, **is** 64. That is the difference between "owing $64" ($-64$ or negative 64) and "having $64" (64 or positive 64).

Observe the context.   Say to yourself: Where does this idea fit into a larger picture? What is going on here overall? What are the circumstances? For example, .02 + .8 + 2.35 looks like money, even if it's inches. You know how to add money!

Observe sights, sounds, or smells.   Even math classes change from day to day.

- Solving problems the day the teacher wore a Hawaiian shirt could be recalled as the "Surfer Procedures."
- Varying background songs when you study will attach what you learn to that song. Humming the song can trigger the memory.

5. Be Active with What You Want to Learn

Manipulate and experiment. Find real-life examples of what you are learning. Make analogies that you understand. Talk about it, read more about it, sketch it, question it, dance with it, rhyme with it, or Mind Map it. Here are other suggestions of activity with learning:

- Work problems on a chalkboard.
- Use the Finger Technique from Chapter 2's "Master Math's Mysteries" to learn the nine multiplication tables.
- Notice geometric shapes and patterns in nature around you.

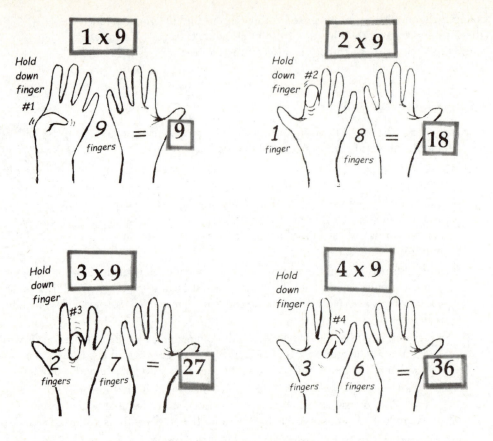

- Use paper or cookie dough to make models of problems.
- Think about distances, rates, and times as you drive, or pretend your eraser and a paper clip are the two cars racing in the problem. Replicate the distances, speeds, and times with actions.

6. Chunk the Information

Research tells us that we can remember only about seven bits of information at a time. Break information down into smaller, easier-to-remember pieces of three to four bits each. This makes it more manageable. For example, when I type student identification numbers into my grading programs, I chunk the nine-digit numbers into three shorter pieces. I think of the student number 463528934 as the three shorter numbers 463, 528, and 934. Another example, found later in this chapter, uses chunking in a mnemonic device I invented to remember the names and locations of 19 streets in my neighborhood.

7. Tune Everything Else Out

Stop interference. Your psychic energy is limited, so focus it on your studies. In class, do not sit by distracting people even if they are your friends. Outside of class, study in a location where your brain gets the privacy it needs.

- **For auditory learners:** Use earplugs or listen to music. Turn down the TV. Stake out a quiet section of the library or the math-tutoring center.

- **For visual learners:**   Clear the visual field in front of you. In class, sit in front so you don't see other students and nonmath action. Use a study carrel in the library. Clear your worktable or desk before you begin. Choose a desk or work area facing a neat and pleasing background.

- **For kinesthetic learners:**   Choose workplaces where you have space to move as you need. Give thought to your physical comfort.

If you do not know whether you are an auditory, visual, or kinesthetic learner, review Chapter 3.

8. Have Fun

Intention and good humor help brain function. When you play and make jokes with your work, you will do better work. Smile. Laugh a lot and enjoy the absurdity of it all. Work with positive people in a study group. Don't sit by negative complainers. Math jokes are usually corny. Even "corn" can give you a good time.

Math Corn

Q: What did the little acorn say when he grew up?

A: Gee-Ah'm-a-tree

Figure 8.3   *Mind Map of Eight Ways to Take Math into Your Brain Memorably*

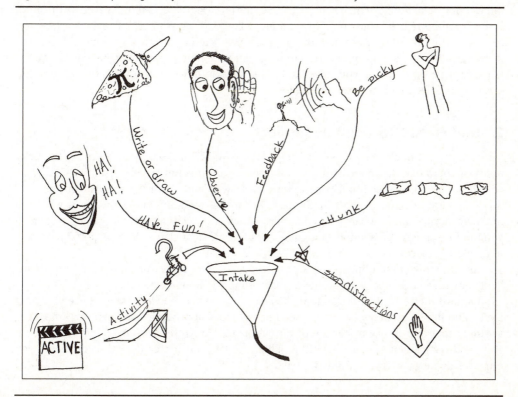

Second Step—Storage: Store Math in Your Mind Solidly and Permanently

Keep those connections! Once you take math processes and ideas into your mind, these six techniques will help your brain to store them solidly and permanently. As you read, check those actions you are ready to take (see Figure 8.4).

 HOW TO STORE MATH IN YOUR MIND SOLIDLY AND PERMANENTLY

1. Say to yourself, "This information is important."
2. Use hard storage for support.
3. Use it not to lose it.
4. Allow yourself settling time.
5. Develop habits.
6. Sleep.

1. Say to Yourself, "This Information Is Important"

Nonessential information disappears quickly. When the instructor points out important information, making a mental note to yourself of its importance solidifies it in your mind. Focus on important information. For example, when parking your car in a large parking lot, recognize the importance of remembering the car's location as you take notice of the sign markings and the view from your car to help you remember later. Recognizing the importance might even motivate you to write the information down or to review the information several times. (Remember: Use it not to lose it!)

2. Use Hard Storage for Support

Write things down. Use a notebook, a notepad, note cards, a journal, personal calendar, sticky notes, or computer files to record and recall essential information. The last part of this chapter will show you how to use Mind Maps to form a visual summary of ideas about a concept.

Take careful notes in your math class. Record all assignments. Copy all of what the instructor writes on the board as well as oral explanations. Develop a note-taking system so you can find assignments, important points, or examples at a glance. See Chapter 5 to review note-taking procedures.

As you study, take notes from your textbook of important vocabulary and problems. Write down key concepts with problems that exemplify those ideas.

Fictional detective Kinsey Milhone in Sue Grafton's O *Is for Outlaw* says, "I used to imagine I could hold it all in my head, but memory has a way of pruning and deleting, eliminating anything that doesn't seem relevant at the moment." So Kinsey uses index cards to record clues and facts about her investigations. Then she lays out her cards and reads them over, often seeing new patterns and solving crimes.

3. Use It Not to Lose It

Repetition and practice reinforce the wiring in your brain. Revisit the concepts and processes within hours of class, and then review again in the following days and weeks. To "use it" try these activities:

- Talk to yourself about math and discuss math ideas with other students. Tutor others. Work problems over again without the aid of your book or notes.
- Work the chapter review problems from the book. Make a list of examples that your teacher presented in class. Work those problems once a week throughout the semester.
- Do three review problems each day as well as the current homework.

4. Allow Yourself Settling Time

Pace yourself. Your brain needs to process math information with no new challenging input coming in. Spend from 20 to 50 minutes learning and then spend about 10 to 15 minutes resting or doing something unrelated to what you just learned. Settling time does not have to last long. It can be as short as walking to your next class, taking a rest-room break, chatting with your neighbor, or stepping outside to get a breath of fresh air. During a math class that lasts over an hour and a half, those breaks are extremely impor-tant. A knowledgeable teacher will build settling time into the class structure. Taking a break to joke around or move the class into groups or to the board allows students the breather they need. I use math songs, silly stories, or relaxation exercises to provide settling time in my classes.

5. Develop Habits

Your habits—what you repeat again and again—will form well-established pathways in your brain. Develop successful routines for common procedures such as note taking, study-ing, or problem solving. For example, form the habit of taking a deep breath every time you begin to work on a word problem. Read the word problem through once just to figure out what you are asked to find. Take a second deep breath, expelling all the air. Then read the problem through two more times before you ever begin working. These habitual activities will relax you and remind you that you are not expected to understand word problems immediately.

Form such study habits as labeling all class notes with the date, name of the class, and page numbers; highlighting important concepts in red; heading to the library right after class to begin studying; and keeping your study tools (book, notebook, pencils) in a special place where you can always find them.

6. Sleep

To remember, you require sufficient rest, especially early morning sleep. The hours between 3 A.M. and 6 A.M. are the hours when you process and understand your long-term memories during rapid eye movement (REM) sleep. Studies show that a lack of sleep interferes with brain function.

Figure 8.4   *Mind Map of Six Ways to Store Math in Your Mind Solidly and Permanently*

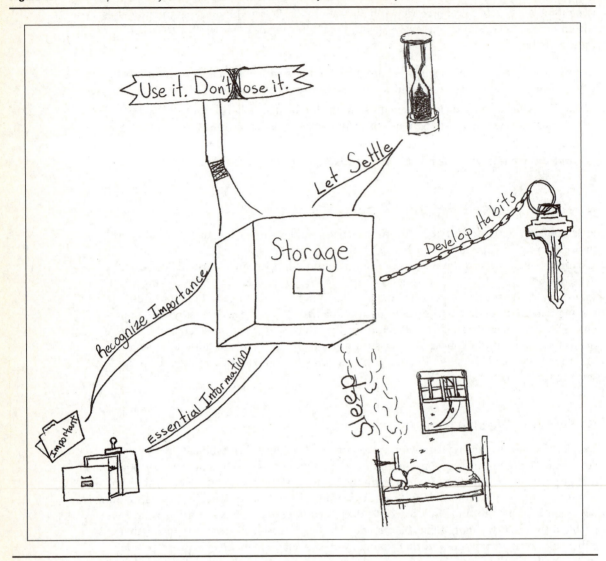

Third Step — Recall: Retrieve the Math You Already Know

Be able to retrieve those brain connections that you have made. Each step for making long-term memories is crucial. If the pathways have not been formed and stored well, the information cannot be reached or remembered.

The key to retrieving or accessing what you have stored in your brain is to understand and to *practice*. The more you practice the math concepts and techniques, the more easily you can remember them as needed. Here are techniques that you can consciously use to recall facts, formulas, procedures, or definitions. These techniques don't explain why a procedure works, but they do help you recall how to do the procedure once you understand it. Understanding greatly helps your recall (see Figure 8.5).

HOW TO RETRIEVE THE MATH YOU ALREADY KNOW

1. Take in and store math carefully.
2. Link actions with words or ideas.
3. Link images with words or ideas.
4. Use mnemonics.

1. Take In and Store Math Carefully

The best way to retrieve the math you already know from your mind is to have taken it in and stored it carefully in the beginning. Reread the strategies earlier in this chapter for the First Step—Intake and the Second Step—Storage, and always use them as you learn.

2. Link Actions with Words or Ideas

Actions create motor memory. Move around as you say the ideas aloud. Create a little dance with them. Have you noticed that you remember certain events as you do certain things? When I do one of my yoga exercises, I always remember a fellow yoga student. My action is linked with my memory of Julie and what she said about the exercise. Linking words or ideas with actions can be used to remember anything and at any math level from basic math into more advanced math. Here are some other examples:

- Flip your right hand over and say "reciprocal." The reciprocal of $\frac{2}{3}$ is $\frac{3}{2}$. The reciprocal of 5 is $\frac{1}{5}$ because 5 is $\frac{5}{1}$. Eventually when you see "reciprocal," your right hand will automatically twitch and you will remember to flip the fraction over to find the reciprocal.

- To remember how to multiply and divide fractions using actions with your right hand, see the "Master Math's Mysteries" in this chapter and in Chapter 9.

- One of my teaching colleagues had her students line dance the shapes of important math graphs. A lot of jumping and slinking up and down made those graphs memorable.

- I memorized a French poem while walking back and forth to class many years ago. Unfortunately, this poem is the only French I remember from three years of studying the language in college. I do not even know what the words mean, but they are inscribed into my brain. The only explanation I have is that the motion formed strong connections in my mind. Hindsight is great. Now I know I could have remembered much more by reciting my French lessons aloud as I walked from my dormitory to class! Try it with your math work.

3. Link Images with Words or Ideas

Vivid pictures involve sensory memory.

- Disneyland knows this technique. The park places images of Disney characters on signs throughout its parking lots to catch the eyes of visitors and remind them of the locations of their cars. (Can you imagine searching the Disneyland parking lot for your lost vehicle?)

- I link images with ideas as I make a little sketch to remember my to-do list of six or seven errands such as pick up dog food, pick up laundry, get money, reserve a hotel room, call Mike, buy orange juice, and gas the car. Each errand appears symbolically in the sketch, which I post on my garage door.

■ Writing the important math words *sum, difference, product,* and *quotient* in the form shown here gives a vivid image of the meaning of each word. Sum is written as a plus sign, difference is written as a subtraction sign, product is written as a multiplication sign, and quotient is written as a division sign.

```
                                    P               T
                                  R       C               Q
                                  O   U
          S                       O   U
          S  U  M   DIFFERENCE    D           UOTIEN
          M                       O   U
                                  R       C           T
                                  P               T
```

■ Ad agencies associate an image with an idea when they show young, beautiful people with their products. They want you to associate youth and beauty with their products and purchase them.

4. Use Mnemonics

Make up sentences, goofy words, songs, and rhymes that state, spell out, or remind you of the first letter of the words you want to remember. The more meaning they have for you, the better you will remember. Making them funny helps, too. The list has examples of mnemonics. The first three are commonly used outside the math world.

- "*Spring* ahead; *fall* back" to remember which way to turn the clock for daylight saving time.

- "*All Cows Eat Grass*" to remember the names A, C, E, and G of the spaces in the bass clef in musical notation.

- "Thirty days hath September, April, June, and November. All the rest have 31 excepting February" to remember the number of days in each month of the year.

- Several mnemonics are illustrated at the end of "Master Math's Mysteries" in Chapter 3 to remember the words *sum, difference, product,* and *quotient.* Sing the song shown there to remember that "sum" means to add, "difference" means "to subtract," "product" means to multiply, and "quotient" means to divide.

- The acronym PEMA in "Master Math's Mysteries" in Chapter 3 recalls the order of operations for basic math: *P*arentheses, *E*xponents, *M*ultiplication and division (from the left to the right), and *A*ddition and subtraction (from the left to the right).

- Jaime Escalante, the East Los Angeles math teacher featured in the movie *Stand and Deliver,* taught his high school students "*All Seniors Turn Crazy*" to help them remember some basic facts for their trigonometry class.

Figure 8.5 *Mind Map of Four Ways to Retrieve the Math You Already Know*

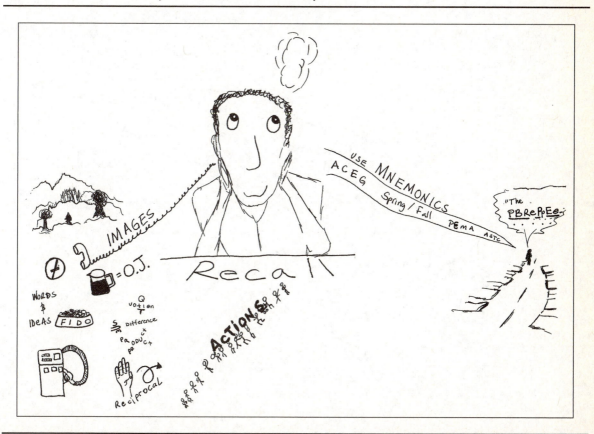

■ I made up the sentence "The *PBRePpEe* from the *ClubhouseS* has *TMJ* and has to *MC* the *CBs* and the *POSsuMS*" to help me remember the order of 19 side streets off a main street where I often walk. I stopped getting lost as I took my morning walk once I memorized this crazy sentence. At first, the sentence began with "The Preppy" but I realized I had left out the street Boa Vista, so I added "B" to the word. The streets are Pitcairn, Boa Vista, Rhodes, Paros, Elba, Clubhouse, Samar, Tahiti, Maui, Java, Mindanao, Country Club, Ceylon, Baker, Palau, Oahu, Samar, Minorca, and Serang. Notice that this example uses chunking as well as a mnemonic device because I grouped the first five streets together in PBRePpEe, the next two streets in ClubhouseS, the next three streets in TMJ, and so on.

Putting the First Step—Intake, Second Step—Storage, and Third Step—Recall altogether, you have a reliable system for developing solid dendritic networks about math in your brain. The Mind Map in Figure 8.6 outlines the steps and techniques.

Figure 8.6   *Mind Map of Three Steps to Solid Memories of Math and How to Achieve Them*

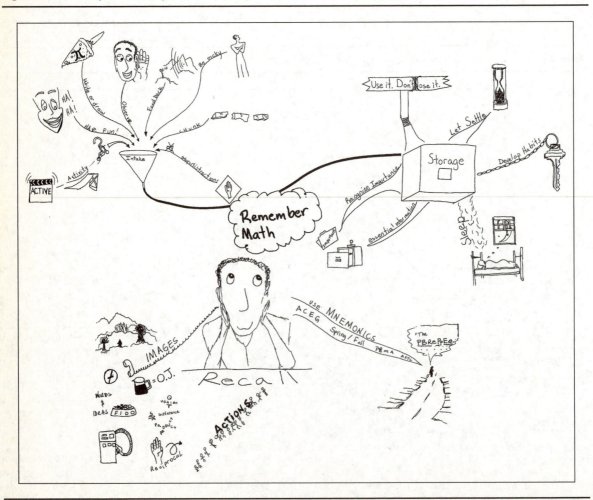

SHORT-TERM GOALS TO FLOW WITH MAKING SOLID MATH MEMORIES IN CLASS

1. Avoid interference. Sit away from friends and distracting people so that you can see and hear only what the teacher is doing.

2. Observe the lesson closely, noticing patterns, similarities and differences, key words, and the context.

3. Get feedback by questioning what you don't understand.

4. Make hard storage by taking notes.

5. Be active during the lesson by making guesses and answering questions. Volunteer often. Mistakes are great! They earn you immediate feedback that polishes your understanding.

6. Laugh and smile often.

7. Label the math material important. Look for main topics and essential concepts.

8. Work all practice problems in class and again after class to use your knowledge instead of losing it.

9. Let your imagination go wild thinking up any actions, images, or mnemonics for today's material.

Mind Mapping®

Mind Maps are powerful tools increasing your ability to remember and synthesize material. They can also be used for reviewing math, preparing for exams, organizing notes, and planning your studies. Also called webs, concept maps, clusters, graphic organizers, and sequential organizers, Mind Maps increase math understanding by connecting and organizing concepts. Mind Mapping uses many of the memory techniques presented earlier in this chapter such as being active, chunking, labeling information important, using hard storage, and linking images with ideas. Mind Mapping can even be done on the computer with one of the many Mind Mapping software packages or free programs on the Web.

Notice how a sequence of Mind Maps in Figures 8.2 to 8.6 earlier in this chapter illustrates the development of the three steps for remembering math. (They were made for you by Santa Ana College math student Carlos Ordiano, who is featured at the end of Chapter 7.) Figure 8.6 summarizes the whole process. See whether you can verbalize the three steps and how to accomplish them. Mind Maps clarify ideas associated with a basic focus. Our focus in Figure 8.6 was "remembering math." However, each of the three steps for remembering math had many subtopics, so each of those three steps became the focus for a separate Mind Map in Figures 8.3 to 8.5. Mind Maps illuminate connections and order, and they make ideas accessible. They take you beyond linear thinking.

The skills involved in making Mind Maps appear in the journals of some of the world's most inventive minds, such as Leonardo da Vinci, Thomas Edison, Albert Einstein, and Charles Darwin. Author and educator Tony Buzan* (1994) linked these skills to brain function and popularized the Mind Mapping technique through training and writing. He says that

---

* For further information on Tony Buzan and Mind Maps,® see www.buzanworld.com.

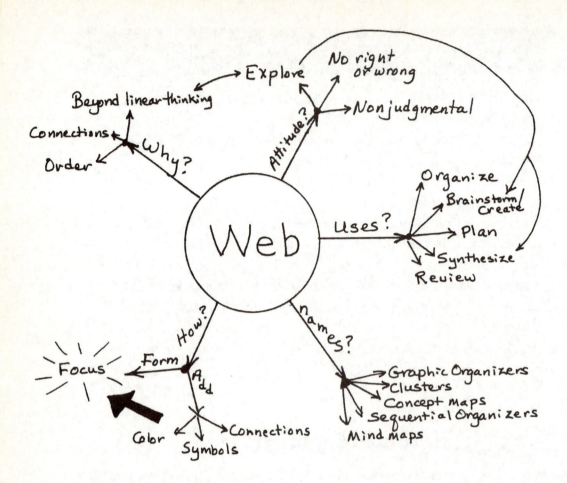

the results of this process mirror the associative thought processes of the brain. His *The Mind Map Book* contains detailed examples and many exciting applications. It is one of my top 10 all-time favorite books.

How to Mind Map

1. In the center of a blank sheet of paper, write the main idea or focus you wish to explore. Preferably use one word or a symbol. Imagine that this is the hub of a bicycle wheel.

2. Surround the main idea with all of the related ideas by drawing lines out like the spokes of a wheel. Print one word or symbol representing each related idea on each line or at the end of it (e.g., Figure 8.2).

3. Elaborate on related ideas. If you wish, add spokes to the related idea, making a miniwheel (e.g., Figures 8.3, 8.4, and 8.5).

4. Add color. Show any other connections. Use pictures, words, and symbols.

Buzan says, "The only barrier to the expression and application of all our mental skills is our knowledge of how to access them" (p. 33). Mind Mapping can help you access more of your mental skills if you are willing to explore and experiment. Remember that everyone thinks differently, so your Mind Maps will not be the same as those of others. You can, however, get ideas from other people's Mind Maps to improve your own. Even if you are uncomfortable, by persevering and using this technique you could potentially break through to a *whole new way of thinking*. Carlos Ordiano found Mind Mapping to be a great way to learn the material in his general education courses—those courses outside his main interest.

Be nonjudgmental as you make your Mind Maps. They are for you alone. There is no right or wrong way to make them. Your lines do not have to be perfectly straight, and your pictures do not have to be masterpieces. You can use any color and any color code you wish. You can edit them as you notice new connections. You can even use the same main focus and start over, making a whole new map.

Redrawing a Mind Map several times, adding details each time, helps you learn the ideas that you want to fix in your mind. Redrawing it is especially helpful when you prepare for an exam, review your course, set the ideas of the course in your mind, or look for new connections in the material. If you need to make a presentation in your math course (or any of your other courses), Mind Mapping the material and then redrawing your map will cement the presentation in your mind. You can even sketch your map on your exam or for your audience to recall details.

I had always considered myself a noncreative, linear thinker until I learned to Mind Map by observing Santa Ana College counselor Dennis Gilmour as he taught my math students in his counseling class. To create this book, I made new Mind Maps constantly as I worked on the manuscript. When I felt either stuck or inspired, I would take 10 minutes to brainstorm all of the ideas I thought should be in the book. Often I started my daily writing by mapping my ideas for the day's topic, such as "Taking math tests." I put that main idea in the center of my journal page and, with stream of consciousness, surrounded it with all of my thoughts about the topic. Then I put lines and arrows between related ideas. Often I would look back at previous maps on the same topic to see whether I had new ideas or connections. Writing came more easily after this activity.

Suggested Uses for Mind Mapping with Math

1. Take notes.   Listening carefully to your math instructor for the organization of the lecture, put the topic of the class day in the center of your paper and fill in different procedures and examples as related ideas. You can circle related ideas and draw in arrows to show connections. Here is a Mind Map of notes on adding fractions. Remember that your thought process is unique, so this Mind Map may make no sense to you. Your own Mind Map would look different.

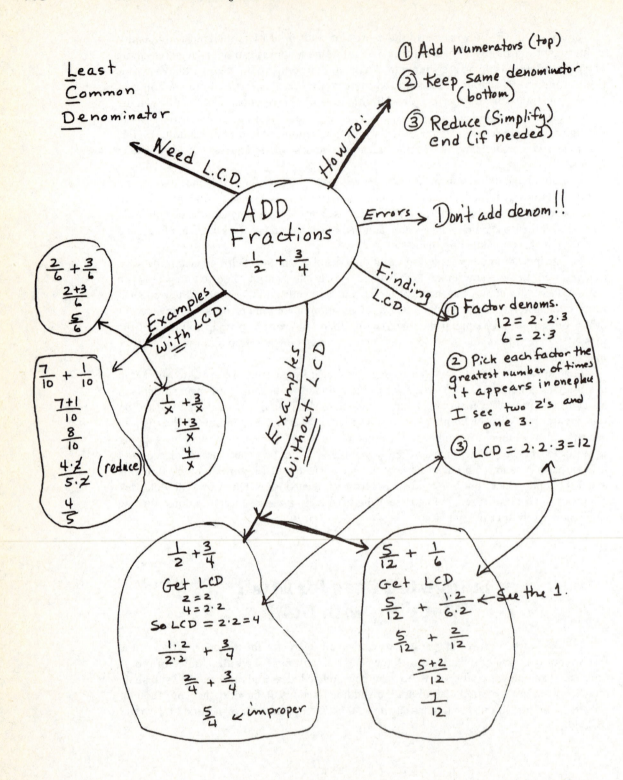

Least
Common
Denominator

① Add numerators (top)
② Keep same denominator (bottom)
③ Reduce (Simplify) end (if needed)

Need L.C.D.

How To:

ADD
Fractions
$\frac{1}{2} + \frac{3}{4}$

Errors → Don't add denom!!

Finding L.C.D.

$\frac{2}{6} + \frac{3}{6}$

$\frac{2+3}{6}$

$\frac{5}{6}$

Examples
With LCD.

$\frac{7}{10} + \frac{1}{10}$

$\frac{7+1}{10}$

$\frac{8}{10}$

$\frac{4 \cdot \cancel{2}}{5 \cdot \cancel{2}}$ (reduce)

$\frac{4}{5}$

$\frac{1}{x} + \frac{3}{x}$

$\frac{1+3}{x}$

$\frac{4}{x}$

Examples
Without LCD

① Factor denoms.
  $12 = 2 \cdot 2 \cdot 3$
  $6 = 2 \cdot 3$

② Pick each factor the greatest number of times it appears in one place
  I see two 2's and one 3.

③ LCD = $2 \cdot 2 \cdot 3 = 12$

$\frac{1}{2} + \frac{3}{4}$

Get LCD
  $2 = 2$
  $4 = 2 \cdot 2$
So LCD $= 2 \cdot 2 = 4$

$\frac{1 \cdot 2}{2 \cdot 2} + \frac{3}{4}$

$\frac{2}{4} + \frac{3}{4}$

$\frac{5}{4}$ ← improper

$\frac{5}{12} + \frac{1}{6}$

Get LCD
$\frac{5}{12} + \frac{1 \cdot 2}{6 \cdot 2}$ ← See the 1.

$\frac{5}{12} + \frac{2}{12}$

$\frac{5+2}{12}$

$\frac{7}{12}$

2. **Review your math course.**   Put the name of the course or concept you want to review in the center as the hub. Add the related ideas surrounding it like the spokes of a bicycle wheel. Here is an example of using this technique to review basic algebra. You do not have to understand algebra to notice that I consider five major ideas/procedures in algebra to be simplify, evaluate, factor, graph, and solve. Chapter 12 explains those vocabulary words.

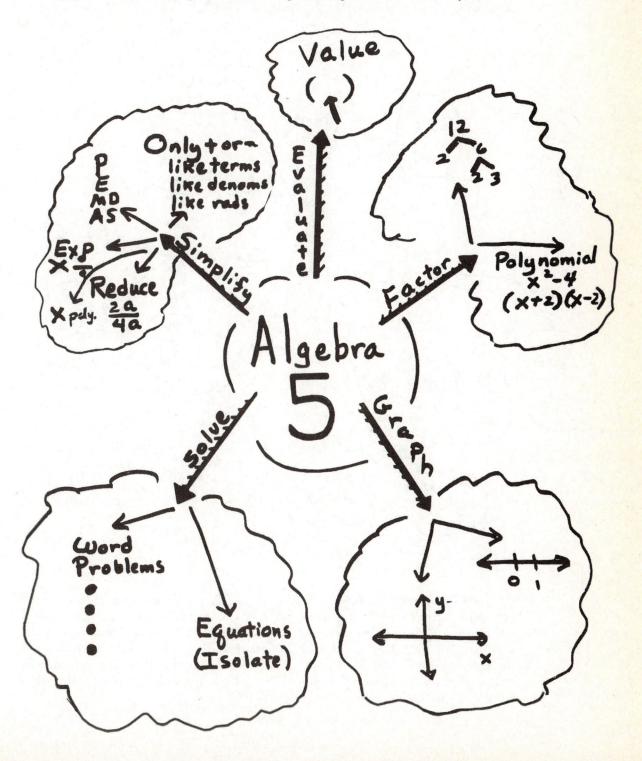

3. Prepare for exams.   Several days before a test, write the words "Next exam" or "Exam #3" or "Chapter 4 exam" in the center of a blank paper. Using your notes and your book, find the main topics that will be covered. Print these topics on the spokes and fill in words and examples appropriately. Placing sample problems from class notes into the Mind Map would be excellent test preparation. Here is an example of a Mind Map for preparing for an exam that would be given early in my prealgebra course.

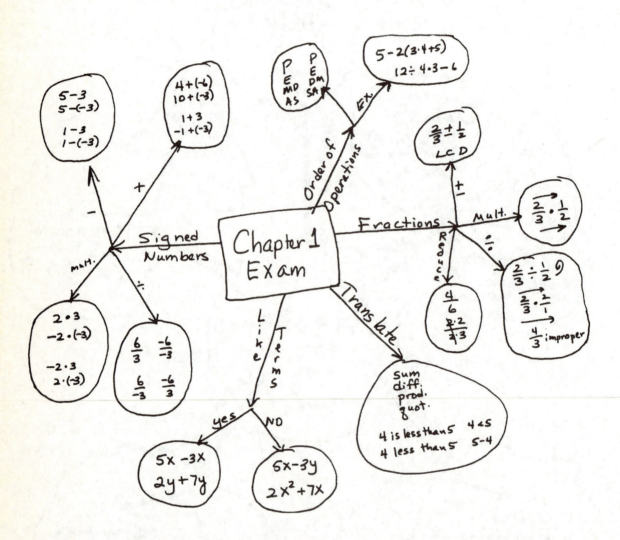

4. Plan your studies.   As the main focus for planning, you could write "To do," "Study activities," "Week's work," "Final exam prep," or whatever you need to plan. Brainstorm activities that need to be done as the spokes and then break each activity down further. Draw in connections. Use highlighters to mark your priorities. Here is an example of a brainstorming beginning map for "Study activities." You can see that many of the subtopics are related and could be placed under one heading. This is only a preliminary map for getting ideas down where you can see them and then organize them.

Study Activities

- Write list of questions
- Visit instructor
- Work problems on blackboard
- Write notes in graphics organizer
- Make flashcards
- Go to math lab
- Make a web of chapter ideas
- Write practice exam
- Work examples from class notes
- Complete homework
- Join study group
- Read book (aloud)
- Find study partner
- List chapter's main ideas
- List hard problems. Do them many times
- Make vocab list

ACT FOR SUCCESS | CHAPTER 8

1. Talk to people about memory skills. Ask whether they have any techniques that have worked for them. Talk about what you remember well and hypothesize about why that is.

2. Write down three ways in which you can improve your recall. Make up personal examples for those techniques.

3. Make up a silly word or sentence to recall an important fact in the math that you are studying right now.

4. Mind Map by writing "Math Success" in the middle of a big sheet of blank paper. In five minutes, surround these words with the actions you associate with math success from reading this book.

5. Mind Map for test prep. Write down "Next Test" in the middle of a big sheet of paper. In the next five minutes, surround these words with the concepts and problems that you believe will be on your next test. Take 10 more minutes to go through your book or class notes, adding and elaborating. Put page numbers or problems by topics. To continue your test preparation, set aside 30 minutes to write down specific problems for each topic. You can find problems in your notes or textbook. Copy the answers on a separate piece of paper. Set aside another 30 minutes to work the problems and correct them. If you are not currently in a math class, do this exercise for any of your classes.

MASTER MATH'S MYSTERIES

Multiplying Fractions and Signed Numbers

Multiplying Fractions

The "why we do it" of multiplying or dividing fractions is more difficult to explain than the "how to do it." I explain the why and how for multiplication in this chapter and for division in Chapter 9.

Review.   Using the Egg Carton Calculator introduced in Chapter 5, recall that each pocket was one twelfth, or $\frac{1}{12}$, of the whole egg carton. Thinking about the 12 pockets, remember how other fractions could be represented like this:

$\frac{1}{6}$ of the egg carton was two pockets, or $\frac{2}{12}$

$\frac{1}{4}$ of the egg carton was three pockets, or $\frac{3}{12}$

$\frac{1}{3}$ of the egg carton was four pockets, or $\frac{4}{12}$

$\frac{1}{2}$ of the egg carton was six pockets, or $\frac{6}{12}$

Multiply Those Fractions.   I offer a few examples of multiplying using the egg carton. It accommodates twelfths, sixths, fourths, thirds, and halves quite well because all of the numbers 12, 6, 4, 3, and 2 divide evenly into 12 or one dozen. With these examples, I hope that you will understand *why* we get the answers that we do. **Hands-on work with egg cartons is**

exactly what many minds need to understand. Both the concrete model and the motion set the ideas in real life. I hope you do what you need to work through and absorb these examples. Use an egg carton to model these problems. Make diagrams of the carton and its pockets shading what is discussed.

a. $\frac{1}{3} \cdot \frac{1}{2}$ This problem can be read "one third times one half" or "one third of one half." Look at one half of the egg carton. That "one half" contains six pockets. If we split that "one half" into three equal parts, two pockets would be in each third. One of those thirds would be two pockets, or two twelfths, or one sixth of the whole carton.

/////////	\\\\\\\\	/////////			
/////////	\\\\\\\\	/////////			

Written out, we have calculated: $\frac{1}{3} \cdot \frac{1}{2}$ is $\frac{1}{3}$ of $\frac{1}{2}$, which becomes $\frac{2}{12}$ or $\frac{1}{6}$

/////////					
/////////					

Conclusion: $\frac{1}{3} \cdot \frac{1}{2} = \frac{1}{6}$

b. $\frac{2}{3} \cdot \frac{1}{2}$ This problem is similar to problem (a). $\frac{2}{3} \cdot \frac{1}{2}$ is $\frac{2}{3}$ of $\frac{1}{2}$. So split one half of the egg carton into three equal pieces. Two of the thirds put together contain four pockets. Those four pockets are four twelfths or one third of the whole carton. So $\frac{2}{3} \cdot \frac{1}{2}$ is two thirds of one half, which is four pockets, or four twelfths, or two sixths, or one third, of the whole carton.

/////////	\\\\\\\\	/////////			
/////////	\\\\\\\\	/////////			

Conclusion: $\frac{2}{3} \cdot \frac{1}{2} = \frac{4}{12} = \frac{2}{6} = \frac{1}{3}$

/////////	\\\\\\\\				
/////////	\\\\\\\\				

(continued)

c. $\frac{1}{6} \cdot \frac{1}{2}$ This problem asks for one sixth of one half. Splitting one half of the egg carton into six equal pieces puts one pocket in each sixth. So one sixth of one half is one twelfth of the whole carton.

Conclusion: $\frac{1}{6} \cdot \frac{1}{2} = \frac{1}{12}$

d. $\frac{1}{4} \cdot \frac{2}{3}$ This problem asks for one fourth of two thirds. From previous work, we know that $\frac{2}{3}$ of the egg carton is eight pockets. Split those eight pockets into four equal pieces, and notice that each piece has two pockets. This means that one fourth of two thirds is two twelfths, or one sixth of the whole carton.

Conclusion: $\frac{1}{4} \cdot \frac{2}{3} = \frac{2}{12} = \frac{1}{6}$

Summarize with a Two-Finger Motion.   Notice that you could do each of the multiplication problems that we worked without the egg carton. The answer can be found by multiplying the tops of the fractions and also multiplying the bottoms of the fractions. Sometimes the answer could be simplified further, but an initial answer comes from multiplying right straight across—top times top and bottom times bottom. Kinesthetic learners or students with bodily-kinesthetic intelligence may want to hold two fingers of their *right* hand pointing straight ahead and move them to the right saying, "Multiply fractions right straight across." Repeating this motion with the speech often reminds students how to multiply fractions correctly.

Distinguish Between Procedures.   Doing the kinesthetic motion with your right hand while saying "Multiply fractions right straight across" helps students distinguish the procedure for multiplying from the procedures for adding, subtracting, and dividing. Only adding and subtracting procedures keep the bottom (denominator) unchanged. The multiplication procedure multiplies straight across, which changes the denominator, as well as the top (numerator). The division procedure changes the denominator too, but in a different way–see Chapter 9's "Master Math's Mysteries."

Review the previous four examples to see that multiplying straight across does give the same answer.

a. $\frac{1}{3} \cdot \frac{1}{2} = \frac{1}{6}$ because $1 \cdot 1$ is 1 and $3 \cdot 2$ is 6.

b. $\frac{2}{3} \cdot \frac{1}{2} = \frac{2}{6}$ because $2 \cdot 1$ is 2 and $3 \cdot 2$ is 6.

$\left(\frac{2}{6}\right.$ is the same as $\frac{1}{3}$, which is the simplified answer.$\left.\right)$

c. $\frac{1}{6} \cdot \frac{1}{2} = \frac{1}{12}$ because $1 \cdot 1$ is 1 and $6 \cdot 2$ is 12.

d. $\frac{1}{4} \cdot \frac{2}{3} = \frac{2}{12}$ because $1 \cdot 2$ is 2 and $4 \cdot 3$ is 12.

$\left(\frac{2}{12}\right.$ is the same as $\frac{1}{6}$, which is the simplified answer.$\left.\right)$

Try These Examples.   Do them two ways–with the egg carton and by multiplying straight across. (Use the two-finger motion as you do them, saying aloud, "Multiply fractions right straight across." Be sure to use your *right* hand.)

1. $\dfrac{1}{3} \cdot \dfrac{3}{12}$ 2. $\dfrac{1}{2} \cdot \dfrac{1}{3}$ 3. $\dfrac{1}{3} \cdot \dfrac{3}{4}$ 4. $\dfrac{1}{5} \cdot \dfrac{5}{6}$

5. $\dfrac{1}{9} \cdot \dfrac{3}{4}$ 6. $\dfrac{2}{3} \cdot \dfrac{3}{4}$ 7. $\dfrac{1}{8} \cdot \dfrac{2}{3}$ 8. $\dfrac{1}{2} \cdot \dfrac{5}{6}$

Multiplying Signed Numbers

When multiplying signed numbers, there are four possible cases. These cases are positive times positive, positive times negative, negative times positive, and negative times negative. The first three cases make sense to most people when they first see the results. The fourth case is sometimes harder to understand, and it is difficult for many math teachers to explain. Read the explanation of each of these four cases. Examine the examples and then fill in the blanks.

Case 1: Positive Times Positive = _____

   Examples:

   a. $2 \cdot 3 = 6$ **b.** $4 \cdot 7 = 28$

   c. $150(6) = 900$ **d.** $\dfrac{1}{2} \cdot 8 = 4$

From these examples, you can see that **a positive number times a positive number is a positive number**.

Case 2: Positive Times Negative = _____
Remember that multiplication is a fast way to add the same number to itself. Also recall from "Master Math Mysteries" in Chapter 7 that when you add dark beans to dark beans, you have only dark beans as a result.

(continued)

Examples:

a. $2 \cdot -4$ is adding two negative 4s. That means $2(-4) = -4 + (-4)$, which is 8 dark beans or -8.
So $2(-4) = -8$.

b. $4 \cdot \left(-\frac{1}{2}\right)$ is adding four negative one halves. $-\frac{1}{2} + \left(-\frac{1}{2}\right) + \left(-\frac{1}{2}\right) + \left(-\frac{1}{2}\right) = -2$
Adding 4 half-dark beans gives us 2 whole dark beans.
So $4\left(-\frac{1}{2}\right) = -\frac{1}{2} + \left(-\frac{1}{2}\right) + \left(-\frac{1}{2}\right) + \left(-\frac{1}{2}\right) = -2$

c. $3(-25)$ is adding three negative twenty-fives.
So $3(-25) = -25 + (-25) + (-25) = -75$

Note that because you are adding a group of negative numbers, you get a negative as the total. From these examples, you can see that **a positive number times a negative number is a negative number.**

Case 3: Negative Times Positive = _____
Order does not matter when you are multiplying. We know that 2 times 3 is equivalent to 3 times 2. Therefore, we can switch the order in the three examples for Case 2 and get the same results.

a. $-4 \cdot 2 = 2 \cdot -4 = -8$

b. $-\frac{1}{2} \cdot 4 = 4 \cdot \left(-\frac{1}{2}\right) = -2$

c. $(-25)3 = 3(-25) = -75$

Case 3 problems are equivalent to the Case 2 problems. Therefore, **a negative number times a positive number is a negative number.**

As mentioned before, most people are comfortable with the first three cases. The results seem natural. The fourth case often takes time for students to believe the result. Belief comes with a feeling of overall "rightness" with how mathematics all fits together. Read Case 4 and fill in the blank.

Case 4: Negative Times Negative = _____
We will use patterns to show how this result occurs. Complete these patterns:

a. $9, 6, 3, 0,$ _____, _____, _____, . . .

b. $-9, -6, -3, 0,$ _____, _____, _____, . . .

In example (a), we are subtracting 3, so the pattern is $9, 6, 3, 0, -3, -6, -9$.
In example (b), we are adding 3, so the pattern is $-9, -6, -3, 0, 3, 6, 9$.

We use these patterns as the solutions for these multiplication problems. Because each column of factors forms a pattern, the solutions need to follow a pattern also. Now look at the following vertical patterns. Follow down the first column and notice that the last problem in the first column is the same as the first problem in the second column because $-3 \cdot 3 = 3 \cdot (-3)$.

$3 \cdot 3 = 9$	$3 \cdot (-3) = -9$
$2 \cdot 3 = 6$	$2 \cdot (-3) = -6$
$1 \cdot 3 = 3$	$1 \cdot (-3) = -3$

$$0 \cdot 3 = 0 \qquad\qquad 0 \cdot (-3) = 0$$
$$-1 \cdot 3 = -3 \qquad\qquad -1 \cdot (-3) = 3$$
$$-2 \cdot 3 = -6 \qquad\qquad -2 \cdot (-3) = 6$$
$$-3 \cdot 3 = -9 \qquad\qquad -3 \cdot (-3) = 9$$

Looking vertically:

- First column: 3, 2, 1, 0, -1, -2, -3. We are decreasing by 1.
- Second column: 3, 3, 3, 3, 3, 3, 3. We are keeping the number 3 constant.
- Third column: 9, 6, 3, 0, -3, -6, -9. We are using a pattern developed earlier.
- Fourth column: 3, 2, 1, 0, -1, -2, -3. We are again decreasing by 1.
- Fifth column: -3, -3, -3, -3, -3, -3, -3. We are keeping the number -3 (negative 3) constant this time.
- Sixth column: -9, -6, -3, 0, 3, 6, 9. We are again using a pattern developed earlier.

Now look at the actual multiplication problems. The first column of multiplication problems begins with a positive times a positive is equal to a positive. The first column ends with problems that show that a negative times a positive is a negative. Recall that order when multiplying doesn't matter, so we switch the order in the last problem of the first column to begin the second column with $3 \cdot (-3)$. We then continue the patterns and discover that a negative number times a negative number is a positive number when we see that $-3(-3) = 9$. Looking at the horizontal and vertical patterns yielded the answer to our question.

A **second explanation for Case 4** can be developed using the idea of "opposite." Recall that taking the opposite of a number changes its sign. The opposite of positive three is negative three. The opposite of negative three is positive three.

Examples:

a. "Opposite 5" is -5. The symbol for opposite is the "$-$" sign. Using it we can say that opposite 5 is -5, which is negative 5 or -5. This can be confusing because "opposite" changed into "negative" and looked the same. Remember the discussion in Chapter 7's "Master Math Mysteries."

b. "Opposite -8" is 8. Written in symbols, opposite $-8 = -(-8) = 8$.

c. "Opposite -2.3" is 2.3. Written in symbols, opposite $-2.3 = -(-2.3) = 2.3$.

Having reviewed the concept of "opposite," we can now apply it to Case 4: Negative Times Negative.

Examples:

a. $(-2) \cdot (-3) = -2 \cdot (-3) = $ opposite $2 \cdot (-3) = $ opposite $(-6) = 6$

b. $-7 \cdot (-4) = $ opposite $7 \cdot (-4) = $ opposite $(-28) = 28$

Again, notice that **a negative number times a negative number is a positive number.**

If these explanations are not enough for you to believe that a negative number times a negative number is a positive number, you are not alone. At this point, take it on faith until you can understand why it is always true, or begin to ask teachers or tutors or other students why this is true until you are convinced.

(continued)

♫ A memory technique for recalling the four cases is to sing the following lyrics to the tune of the song "If You're Happy and You Know It, Clap Your Hands." ♪

Multiplying Signed Numbers ♪

A minus times a minus is a plus.	[Case 4]
A plus times a plus is a plus.	[Case 1]
A minus times a plus	[Case 3]
Or a plus times a minus	[Case 2]
Is a minus, yes, a minus,	
Not a plus.	

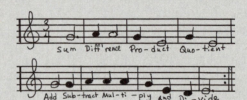

Try these:

9. $-8(-10) =$ _____

10. $11 \cdot (-5) =$ _____

11. $6(-3) =$ _____

12. $-2(20) =$ _____

13. $-40(-25) =$ _____

14. $10\left(-\dfrac{1}{2}\right) =$ _____

Student Success Story

Joel Sheldon

Student **Joel Sheldon**, a bright, creative person who thinks outside the box, says he did a "whole lot of nothing" looking for a good time and working menial jobs in and after high school. In his mid-20s, he found that "good time" by working in the math classroom as a teaching assistant and in the Math Study Center tutoring students from basic math through calculus. Joel remained on the fence about whether to major in mathematics or physics, but he was definite about his love of both. His easygoing, straightforward style is popular with the students he assists. Joel likes the fact that "math can be applied to everything that you see around you." He says, "Knowing that I can do it and that I can convey it to others is one of the things that helped me to like it more." Today, Joel has earned both a bachelor's degree and a master's degree in mathematics and is teaching at the community college level.

CHAPTER 9

Solve Problems Systematically

"It's not that I'm so smart, it's just that
I stay with problems longer."

ALBERT EINSTEIN

Problem solving is a process. It is what you do when you *don't know* what to do. It requires living with uncertainty and trusting that your efforts may work out. As the Number Devil character in Hans Enzenberger's book *The Number Devil: A Mathematical Adventure* (1997) told young student Robert in his dream, "We've racked our brains over this [problem] for quite some time now, and, as I say, the cleverest number devils have tried every trick in the book. Sometimes we can work it out and sometimes we can't."

In this chapter you will practice some of the "tricks" to which the Number Devil was referring. These tricks are really strategies that experienced problem solvers know how to use. There are many strategies, and the more you experience and practice, the more problems you can approach systematically and solve. But before we discuss strategies, I would like to tell you a few stories about problems being solved.

"A great discovery solves a great problem but there is a grain of discovery in the solution of any problem."

Mathematician George Polya (1988, p. v)

Story #1: Fly Watch

My favorite math story is about **René Descartes** lying on his bed as a sickly teenager watching a fly walk the ceiling. In a flash, he recognized that the fly's position could be described with two numbers—the perpendicular distances from the fly to two edges of the ceiling. With this basic idea, he eventually developed the rectangular coordinate system for graphing relationships. *Not while concentrating but while daydreaming,* Descartes discovered the fundamental concept that allows algebraic symbols to be visualized geometrically. He became one of the world's most influential scientists.

Story #2: The Boys' Room

I hurried into the women's restroom at Dodger Stadium during an exciting baseball game. The restroom had two sides of stalls with a bank of sinks in the center. Just inside the door, a small boy wailed to his unhappy mother, "I want to go to the boys' room." The young mother looked frustrated and helpless standing over the sobbing child, who was too young to visit the men's restroom alone. While they interacted, I noticed an available stall in the corner. No women were washing their hands nearby. A thought popped into my mind. I leaned over, pointing to the empty stall, and said to the boy, "There's the boys' room." He stopped crying and ran to the open door, followed by his clearly relieved mother. My "off the wall" white lie temporarily converted the gender of one Dodger bathroom, but it solved a rather loud and urgent problem.

Story #3: Karl Friedrich Gauss

Mathematician Karl Friedrich Gauss (1777–1855) was young when he and his unruly classmates were assigned busywork to keep them out of mischief. They were told to add the counting numbers from 1 to 100 (that is, add $1 + 2 + 3 + 4 + 5 + \ldots + 98 + 99 + 100$). Imagine the teacher's amazement (and perhaps dismay) when Gauss found the correct answer, 5,050, in a matter of minutes. Asked to explain, Gauss said he paired the numbers. Beginning with first and last, $1 + 100 = 101$, then $2 + 99 = 101$, then $3 + 98 = 101, \ldots$, then $49 + 52 = 101$, and finally $50 + 51 = 101$. This makes 50 pairs of numbers adding to 101. Then 50 times 101 makes 5,050. How Gauss decided to do that, we don't know. Maybe he noticed a pattern from staring at the numbers. Maybe he had done a similar problem before. Certainly he would *not* have found this ingenious shortcut if he had obediently begun adding $1 + 2 + 3 + 4 + 5 + 6 + 7 + \ldots$.

These three stories illustrate that some problems and their solutions are important for all humankind. And some problems are only important for those involved. Also, some solutions seem "off the wall" and probably aren't. Other solutions truly are off the wall but are inconsequential in the long run. You and I will most likely not solve the really big problems, but we will solve many problems and we can learn some tricks or strategies from the problem-solving processes of others. As Stanford mathematics professor George Polya says, "[T]here is a grain of discovery in the solution of any problem."

Although the really big problems sometimes seem like they were solved instantaneously, they were not. You will find that a lot of intense study happened first as the problem solver absorbed information and worked with ideas. **It is the curiosity and tenacity of the problem solver, along with direction and intention, that keeps the problem-solving process moving along.** Apparent speed comes only when the solver has experienced the same type of problem or considered the problem in detail previously.

Polya's Four Steps

In 1945, George Polya wrote a little book, *How to Solve It,* that has remained a best seller. He broke problem solving into four amazingly simple, but powerful, steps:

1. Understand the problem.
2. Devise a plan.
3. Carry out the plan.
4. Look back.

Polya's steps can be applied to any problem. Unfortunately many people ignore Steps 1, 2, and 4. They want an answer now, and they carry out the first plan, any plan, they see. In Story #3, Gauss clearly did not skip steps 1 and 2. He surveyed the numbers and noticed a pattern of number pairs adding to 101 instead of just laboriously adding the numbers as the teacher assigned.

When emphasis is on speed, many problems don't get solved. **Bottom line: Solving new problems takes time. What do you do during that time?**

In this chapter, the methods, not the problems themselves, are important. How problems are solved should be your focus. For further practice, look into excellent books such as *The Only Math Book You'll Ever Need,* by Stanley Kogelman and Barbara R. Heller, and *Problem Solving Strategies: Crossing the River with Dogs,* by Ted Herr and Ken Johnson.

Problem-Solving Strategies to Practice

There is no "one best way" to solve problems. There are simply many useful methods that *might* work. The more strategies that you practice, the larger your "bag of available problem-solving tricks." Often many of these strategies work together especially if you are willing to spend time doing Polya's Steps 1 and 2, just becoming familiar with the problem rather than trying to do Step 3 immediately. Gauss (Story #3) most likely discovered his great shortcut to adding 1 through 100 by not jumping into the problem immediately.

"Tricks" that Gauss may have used are listed in the nearby box. The more you practice these strategies, the more they become habits that you automatically do when you approach a new problem. Many of these strategies (State the goal, Decide knowns and unknowns, Throw out irrelevant information, Organize the information) will work on every problem in the beginning. "Consider answers that make sense" and "Confirm your answer" work well for every problem at the end. The other strategies are suggestions to try. Sometimes they work and sometimes they don't. Story #9, "The Handshake Problem" (in the last half of this chapter), tells the story of the most exciting problem-solving adventure I have participated in over my years of teaching math at all levels. The story illustrates nearly all of the strategies listed here, and the problem was always solved by students who were at the basic levels of math and described themselves as "math anxious."

The strategies are described in detail. Use them for your reference now and later when you need some "tricks" to unlock a problem. Notice that most of these strategies (the first 12) would be part of Polya's Step #1, "Understand the problem," or Step #2, "Devise a plan." The last four strategies would be part of Polya's Step #4, "Look back." **Unfortunately most of us jump immediately into Step #3, "Carry out the plan," when we don't really even have a plan. Spending more time on Steps 1, 2, and 4 would help all of us solve many more problems.**

TRICKS TO TRY: PROBLEM-SOLVING STRATEGIES THAT WORK

State the goal specifically.	Decide knowns and unknowns.
Throw out irrelevant information.	Try something "off the wall."
Guess and check.	Simplify the problem.
Act it out or use objects.	Make a picture or diagram.
Organize the data.	Identify patterns.
Work backward.	Use algebra. (See Chapter 12.)
Consider answers that make sense.	Solve the problem another way.
Generalize.	Confirm your answer.

State the Goal Specifically

Get a focus. Write down what you are asked to find. Define what you seek. How will you measure or recognize it? Often the goal or request is stated at either the beginning or the ending of a problem. To state the goal, simply write "Find . . ." and complete the sentence, such as "Find the speed in mph."

Decide Knowns and Unknowns

What does the problem tell you specifically? What don't you know yet? Writing down this information also helps you focus and understand.

Throw Out Irrelevant Information

Which information has nothing to do with what you are asked to find? Eliminate it and use the rest. The problem may give irrelevant details, such as the color of a car or the gender of the people. Even numbers in the problem may have nothing to do with what you are requested to find.

Try Something "Off the Wall"

Brainstorm. Color outside the lines. Hunches or intuitions or wild guesses may seem risky, *but* they may break through to a solution if you work with them. In the beginning, mathematics was created from intuitive guesses and dreaming. The logic binding it all together came later—sometimes hundreds of years later. Think differently from a new perspective.

Guess and Check

Estimate and test your estimate. This process is the basis for most science. Make an easy first estimate and see if it works in the problem. Record your estimate. (Of course, it will be wrong. It is only a guess.) Notice whether the first estimate is too large or too small. Refine your second estimate accordingly. Continue refining each estimate, testing it to see how far off you are. You might be surprised how quickly you can solve your problem. An added benefit to checking many guesses will be your increased understanding of the relationships between elements of the problem, which could help you set up a solution by an alternative method—perhaps by algebra (see Chapter 12).

Simplify the Problem

If possible, bring the problem level down to one case or a few cases. Solve the simpler problem and then work up one step at a time, watching for patterns. Suppose that you are asked to find how many trips 50 people make. First, think about one person, then two people, and so on.

Act It Out or Use Objects

Model the problem. For new perspectives, move objects or yourself around to develop a visual, kinesthetic image of the situation. Use a little drama. If a problem has motion in it, simulate the motion so that you clearly see what happens. Einstein reportedly came up with his famous formula $e = mc^2$ by visualizing himself running alongside a beam of light. Sometimes he pictured himself running alongside and sometimes riding the beam. Einstein experimented with this visualization for years before breaking through to understanding. Are you willing to be that persistent?

Make a Picture or Diagram

Use pencil and scratch paper to draw what is happening. Find a way to sketch the relationships. Organize the details graphically. The results do not have to be artistic, but you do want to be accurate and detailed. You are merely making a visual simulation of the problem, and detail helps you understand. You might make several successive pictures, refining each one as you understand more detail.

Organize the Data

Using a t-table or a chart, list the information in an orderly manner as you discover it. Arranging the data systematically shows you what you know and don't know. Tables help you see any developing patterns and ensure that you have considered all of the possible cases.

Example of a t-table.   Figure 9.1 shows a t-table. The first column is the number of people at a party, and the second column is the number of gifts. The information in this t-table is unorganized. If we organize the information sequentially, we get Figure 9.2 and might begin to see a pattern—that the number of gifts at the party was always two times the number of people.

Figure 9.1   *An Unorganized T-Table*

# of people	# of gifts
2	4
5	10
4	8
3	6
1	2
0	0

Figure 9.2   *An Organized T-Table*

# of people	# of gifts
0	0
1	2
2	4
3	6
4	8
5	10

Identify Patterns

As you might have already noticed, when the data are organized, patterns are easier to identify. Notice any pattern—relevant to the problem or not—but do not get invested in keeping a particular pattern. Let go of anything that doesn't fit the total picture. If Figure 9.2 also showed that 6 people meant 20 gifts, the pattern mentioned earlier would no longer work. A t-table that illuminates patterns helps you get more data or generalize the relationships into a formula.

Work Backward

Starting at the end of a sequence of events and working back in time cracks some problems wide open. This strategy also presents the data from a different perspective and opens your mind to new thoughts about the problem. !DRAWKCAB KROW

Use Algebra

Representing the unknown (what you don't know) with a letter and manipulating symbols solves a wealth of problems. Chapter 12 illustrates The Driving Problem, worked first by guess and check, and then by using algebra.

Consider Answers That Make Sense

What would be too much? Too big? Too small? Too unreasonable? For example, if you are trying to find the speed of a passenger car, an answer of 200 mph makes no sense.

Solve the Problem Another Way

There are many ways to solve a problem. Speculate to find other ways to check the work. Survey other people to look for new methods to solve your problem. You may notice that Story #9 uses more than one strategy.

Generalize

Try to make a general rule that can be applied to similar problems even though the specific details change. Generalizing is what a formula does. Identifying patterns assists in finding formulas. $A = lw$ is a formula that says, "The area of a rectangle is equal to the length of the rectangle times its width."

Confirm Your Answer

Go back over the problem and see how the solution and attempted solutions work. Notice anything new? Do your answers make sense? If your answer is $5.00, does $5.00 work in the original problem?

The Right Answer

Even when we solve a problem correctly, we may not give our answer in a useful form. Answers are determined by what is asked. Always make sure your solution makes sense and exactly answers the question asked. Read the following story and notice how the answer can change with the question asked even when the solving procedure (dividing 13 by 4) is the same.

Story #4: Josh's Class

Five-year-old Josh Parker walks daily with his mom and 2-year-old sister, Claire, from his home to the bus stop by the library. Boarding the school bus with his new friends, he rides up the mountain to his kindergarten classroom of 13 students at Mount Baldy School. Can you solve the following five problems for Josh? They all seem to be solved by dividing 13 by 4, but are the answers "$3\frac{1}{4}$" or "3.25" or "3 remainder 1" or "3" or "4" or "It depends"? Match these answers with the following problems.

    a. If the whole kindergarten class is going to visit the Trout Pond and the school bus is unavailable, how many cars are needed if each parent's car holds a maximum of 4 children?

    b. Josh's principal needs to write a report about field trips. What would be the average number of children per car going to the Trout Pond if 13 children rode in 4 cars?

    c. In the classroom, how would Josh's teachers make groups of 4 with the 13 kindergartners?

    d. One day Josh is given baseball cards by his grandmother. How would he divide 13 baseball cards among 4 of his friends?

    e. If Josh and his three buddies need $13 to purchase a special birthday gift for another friend, how much does each child owe?

Notice all of the different answers to dividing 13 by 4.

    Solution a: Four cars would be needed for the trip to the Trout Pond. Twelve children would fit in 3 cars and 1 more car would be needed for the 13th child. Rounding the answer to 3 would leave one child behind on field trip day. In this situation, the answer to $\frac{13}{4}$ is 4.

    Solution b: The average number for the principal's report is $\frac{13}{4} = 3\frac{1}{4}$ children per car.

    Solution c: Group organization within the classroom would have to depend on the wishes of the teachers because 13 children cannot be split evenly into groups of 4. The teachers could group the children 4, 4, 5 or 3, 3, 3, 4, or ? Thus $\frac{13}{4}$ would depend on teacher judgment and would vary.

    Solution d: Each friend would get $\frac{13}{4} = 3$ cards per boy with one extra to share or give to one boy. So the answer to $\frac{13}{4}$ is 3 remainder 1 because no one would tear a baseball card into four pieces.

Solution e: Each boy would pay $\frac{\$13}{4\text{ boys}} = \$3.25$ per boy for the gift.

As you can see from these answers, "13 divided by 4" does not have just one answer. Finding the right answer that matches the question could keep a valuable baseball card from being cut into four pieces or one child from being left behind on a trip to the Trout Pond.

Stories to Ponder for Now

Here are four situations for your reading pleasure and experimentation. You may wish to pick one and see if any of the strategies from your new bag of tricks could be applied. Solutions to Stories #5–8 are in the Appendix. If you are not ready to jump in now to practice the strategies, skip to Story #9 and read how the strategies are used by my class to solve The Handshake Problem.

Story #5: Pets

Claire had dogs, cats, and birds for pets. If all her pets together had 13 heads and 36 legs, how many birds did Claire have?

Story #6: The Diaz House

The Diaz family just moved into an historic housing district. They are thrilled with their "new" old home and are working to restore it. Mrs. Diaz wants crown molding around the edges of the ceiling in her bedroom, which is 11 feet by 13 feet. Her budget for this project is $200. If the materials and labor for crown molding cost $10 per yard, can she afford this upgrade?

Story #7: Mowing the Lawn

Amy mowed her rich old uncle's lawn. When she had finished, he said, "Amy, I will either pay you $100 today or I will pay you one penny today and double it each day for a whole month. Which do you prefer?" Now, Amy had no pocket money at the moment, and she very much wanted to go to the movies with her friends, who were waiting nearby. They were absolutely dumbfounded when Amy said loud and clear, "I'll take the penny." Why would she do that?

Story #8: Baking Cookies

Emilio baked cookies for his family and friends. He burned the first dozen cookies and threw them out. Then he put away half of what remained for his younger brothers. Half of what was left he gave to his friend Julia. Afterward he ate half of what remained. If Emilio ended with six cookies, how many did he bake?

A Problem-Solving Adventure

The following story about my math students (names changed) is a long, but exciting, story. Reading it will show you how students at the basic levels of math who consider themselves "math anxious" used the problem-solving "tricks" or strategies listed in this chapter to solve a difficult problem for their math level with a few suggestions from me as the teacher. In over 20 years in the classroom, I have presented this problem every semester and every class has risen to the occasion, surprising themselves with their own keen problem-solving abilities. Read on to see how they accomplished this. I thank Cal State Long Beach math Professor Emeritus Ruth Afflack for her permission to both use her ideas and to present this story to you. This story might be more interesting to you if you read it aloud.

Story #9: The Handshake Problem

Math 030 at Santa Ana College is called "Coping with Math Anxiety." One day I walked into this class of 10 students and said that a political science teacher on campus needed our help. I began to read from a paper I had brought with me:

"A political science teacher is going to conduct a workshop for 100 political candidates called "Running for Office: Practical Skills." Her students will be candidates for judgeships, state legislature, county supervisor, and various board positions. They will include 48 men and 52 women. These candidates from two political parties are going to learn, among other things, how to shake many hands without fatigue. So that the political science teacher can plan her time, she needs to know how many handshakes there will be if every candidate shakes every other candidate's hand."

My class looked at me as if to say, "Is this for real?" so I said, "Today this problem is not important to remember but the strategies that we use on it are important. Would you like me to repeat our problem?" Several students said "yes" and I did. Next I asked if anyone would like to see the problem. Three visual learners, Donato, Maria, and José, raised their hands and I passed them my paper.

After clarifying that we only had to tell the teacher the total number of handshakes, not the time allotment, we began to make guesses, which I wrote on the side board. Donna said that 48 and 52 might be included in the answer, but Maria said it made no difference whether the handshakers were men or women. Raul raised his hand and said, "I think it's 99 because you wouldn't shake your own hand." "No," Teresa said, "I think it's 100 × 99." Raul said, "Why?" and Teresa replied, "Because everybody—100 people—will shake 99 hands."

I asked the class, "How will we know who is correct?" Everyone shrugged, so I suggested, "We could bring in 100 people from campus and count their handshakes." There was laughter. Maria, always thinking, said, "We could just count the handshakes among the 10 students in our class and multiply by 10."

I said, "That would simplify the problem, but sometimes it is best to *really* simplify. Perhaps we could begin with 1 person in the workshop. Do I have a volunteer?" After some hesitation, Raul came to the front of the classroom and I said to the class, "Suppose only one person came to the workshop. How many handshakes would you see?" Everyone looked skeptical, but someone blurted out "None!" The others agreed, so I drew a t-table with two columns on the front board, labeling the left column "# of participants" and the right column "# of handshakes." I wrote "1" on the left and "0" on the right.

I turned to the class, saying, "Suppose two people came to the workshop? Donato, would you join Raul to make a workshop of two so we can count the handshakes?" He came to the front and shook Raul's hand saying, "One handshake." The class agreed and I wrote "2" in the "# of participants" column on the left and "1" to the right of it. Then I wrote a "3" in the "participants" column asking for another volunteer.

# of participants	# of handshakes
1	0
2	1
3	?

Elizabeth came forward, and the three students began shaking hands. Quickly they announced "Three handshakes." They shook and counted so fast that we asked to see it again. They were correct. I entered "3" handshakes for 3 participants and wrote "4" in the "participants" column.

# of participants	# of handshakes
1	0
2	1
3	3
4	?

I asked, "Do you have any guesses about the number of handshakes for four participants?" There were guesses of 5, 6, and 8. Then Ilse joined the others to make a workshop of 4 so that we could check. Ilse shook everyone's hand counting "1, 2, 3" then she stepped back. Elizabeth shook Donato's and Raul's hands counting "4, 5." Then Donato and Raul realized that they needed to shake, making 6 total handshakes for the 4 of them. When all the students agreed, I entered 6 handshakes into the table beside 4 participants.

Continuing to increase the number of participants, I placed a 5 in the table on the board. I again asked for guesses about the number of handshakes. Teresa immediately said, "10." Other guesses were 9 and 8, but Teresa looked determined. I asked her if she could explain her guess. She said that the four people already in front of the room had 6 handshakes among themselves and that a new fifth person would add 4 more handshakes, making 10 total handshakes. Teresa came to the front to demonstrate. After the five handshakers shook and counted several times, the class was convinced that Teresa was correct, so we filled in the table and looked at it.

I asked how many handshakes there would be for "6" participants. Several students said, "15." Bob volunteered that the sixth person would add 5 more handshakes to the 10 handshakes already shaken by the 5 participants. In fact, Bob said that adding the left-hand number and the right-hand number in each row of the table gave the right-hand number in the next row down. He came forward to point to the rows in the table and to add. He said, "Starting from the top: $1 + 0 = 1, 2 + 1 = 3, 3 + 3 = 6, 4 + 6 = 10, 5 + 10 = 15$."

# of participants		# of handshakes	
1	+	0	$= \downarrow$
2	+	1	$= \downarrow$
3	+	3	$= \downarrow$
4	+	6	$= \downarrow$
5	+	10	$= \downarrow$
6		15	

The whole class decided that the volunteer handshakers could be seated because they were certain that we could now accurately continue the table without them. I continued the left-hand column through 12. Using Bob's pattern, the class added $6 + 15 = 21$, $7 + 21 = 28$, $8 + 28 = 36$, $9 + 36 = 45$, $10 + 45 = 55$, and $11 + 55 = 66$.

# of participants		# of handshakes	
1	+	0	$= \downarrow$
2	+	1	$= \downarrow$
3	+	3	$= \downarrow$
4	+	6	$= \downarrow$
5	+	10	$= \downarrow$
6		15	
7		21	
8		28	
9		36	
10		45	
11		55	
12		66	
. . .		. . .	
100		???	

I suggested to the class that we could find the number of handshakes by 100 participants if we extended our table continuing to add all the way to 100 participants. The class agreed that it could be done but thought that would take a long time. We decided to look for something else—hopefully a shorter way. I suggested looking for patterns like Bob did, and I told the class that there were many patterns in the number pairs in our table.

Ilse ventured that the answer could be 450. When asked why, she said, "In the table, 10 participants would make 45 handshakes so 10×10 participants should make 10×45 handshakes." I reminded the students that Maria had suggested that strategy in the very beginning, and I explained to the class that Maria's and Ilse's plan would be similar to putting groups of 10 people into 10 rooms and asking them to shake hands within their own rooms. I asked, "Would that be the same as 100 people in one big workshop shaking hands?" Everyone thought awhile and finally Ilse said, "People would only shake hands with people in their own room, not everyone." She withdrew her guess and said that 450 handshakes would be too few. Maria agreed.

We focused again on our t-table on the front board. I repeated, "Bob found a useful pattern and there are others. Any idea you have is worth considering." The room was quiet as everyone stared at the t-table. Elizabeth noticed that the left column was just the counting numbers 1, 2, 3, 4, 5, . . . but that the right-hand column was adding the next counting number every time. She demonstrated by writing the following in the right-hand column of the table:

# of participants		# of handshakes
1		$0 = 0$
2		$1 = 0 + 1$
3	+	$3 = 0 + 1 + 2$
4	+	$6 = 0 + 1 + 2 + 3$
5	+	$10 = 0 + 1 + 2 + 3 + 4$

(continued)

6	$15 = 0 + 1 + 2 + 3 + 4 + 5$
7	21
8	28
9	36
10	45
11	55
12	66
. . .	. . .
100	???

Elizabeth hesitantly said that 100 participants would make $0 + 1 + 2 + 3 + 4 + \ldots + 99$ handshakes—however many that would be. Maria interrupted excitedly that she saw something about multiplication, not adding, and at least it worked for every other pair of numbers. She wasn't sure that it was a pattern, but taking the left-hand number times a counting number seemed to make the right-hand number—sometimes. Maria pointed to the pairs she was considering, and we wrote the following in the table extending her idea to alternating pairs of numbers in the table.

# of participants	# of handshakes
1	$0 = 1 \cdot 0$
2	1
3	$3 = 3 \cdot 1$
4	6
5	$10 = 5 \cdot 2$
6	15
7	$21 = 7 \cdot 3$
8	28
9	$36 = 9 \cdot 4$
10	45
11	$55 = 11 \cdot 5$
12	66
. . .	. . .
100	???

Maria's pattern definitely worked for every other pair. I told the class, "If this is an important pattern, it must work for all the pairs." After looking at the "in-between pairs," we discovered we had to use "halves" in order to multiply the number of participants. (See the next table.) The right-hand multiplier appeared to be increasing by one half from one line to the next.

# of participants	# of handshakes
1	$0 = 1 \cdot 0$
2	$1 = 2 \cdot \dfrac{1}{2}$
3	$3 = 3 \cdot 1$

(continued)

4	$6 = 4 \cdot 1\frac{1}{2}$
5	$10 = 5 \cdot 2$
6	$15 = 6 \cdot 2\frac{1}{2}$
7	$21 = 7 \cdot 3$
8	$28 = 8 \cdot 3\frac{1}{2}$
9	$36 = 9 \cdot 4$
10	$45 = 10 \cdot 4\frac{1}{2}$
11	$55 = 11 \cdot 5$
. . .	. . .
100	???

Next I said that it is easier to spot patterns when the numbers are in the same form, and I suggested we convert all of the right-hand multipliers to "halves." The table began to look like this:

# of participants	# of handshakes
1	$0 = 1 \cdot \frac{0}{2}$
2	$1 = 2 \cdot \frac{1}{2}$
3	$3 = 3 \cdot \frac{2}{2}$
4	$6 = 4 \cdot \frac{3}{2}$
5	$10 = 5 \cdot \frac{4}{2}$
6	$15 = 6 \cdot \frac{5}{2}$
7	$21 = 7 \cdot \frac{6}{2}$
8	$28 = 8 \cdot \frac{7}{2}$
9	$36 = 9 \cdot \frac{8}{2}$
10	$45 = 10 \cdot \frac{9}{2}$
11	$55 = 11 \cdot \frac{10}{2}$
. . .	. . .
100	???

Again we viewed our table, seeking patterns. Excitement seemed to pass through the room, and several students seemed to take in a breath all at once. Jorge and Donna spoke at the same time, saying, "100 times 99 divided by 2." I asked, "What are you seeing?" Jorge explained that he noticed every right-hand number was found by multiplying the left-hand number by 1 less than itself, then dividing by 2. He said, "For example, 10 participants would make 10 times 9 divided by 2 handshakes. Eleven participants would make 11 times 10 divided by 2 handshakes." His explanation began to sink in and many students smiled, saying, "A-ha. 100 times 99 divided by 2." Ilse grabbed her calculator and others grabbed their pencils and paper, calculating an answer of 4,950 handshakes. Teresa, who initially had offered "100 × 99" as a solution, said, "I was almost right. The number of handshakes by 100 people is '100 × 99 divided by 2.'"

I asked "What if there were 200 participants? How many handshakes?" Donna volunteered "200 × 199 divided by 2." I pointed out that we had found the pattern that could make a formula to give us the number of handshakes for any number of participants. "We could go on the road and offer assistance to all political science teachers with their workshops." The class laughed.

We looked at the numbers that Raul and Teresa had guessed at the beginning. Raul guessed "99" because everyone would shake 99 hands and we saw that the number 99 was definitely an important part of the solution. Teresa had guessed "100 × 99," which was very close except that she did not divide by 2. When Teresa wondered why we had to divide by 2, I asked her to shake my hand saying, "How many handshakes do you see?" She said, "One." I said, "Not dividing by 2 counts this one handshake as one for you and one for me. Since this is only one handshake, '100 × 99' must be divided by 2 because it took 2 people to make one handshake."

I reminded the class that Elizabeth had also figured that 100 participants would make 1 + 2 + 3 + . . . + 97 + 98 + 99 handshakes. We had ignored her idea and gone in another direction when Maria discovered the multiplication pattern. Students wondered if 1 + 2 + 3 + . . . + 97 + 98 + 99 would be the answer 4,950 that we now knew to be correct. I left that question in the air for consideration for the next class.

I had noticed Jorge and Donna marking on a shared paper. I asked to look at their paper. They had drawn a series of dots connected by lines that look like this:

When they drew their diagrams on the board, the others agreed that they had drawn a model of the workshops that we had acted out. Dots represented the participants and line segments represented handshakes. Making a t-table from their diagrams exactly matched our t-table.

# of dots	# of line segments
1	0
2	1
3	3
4	6
5	10

I pointed out how many students who thought that they could not solve math word problems participated in the solution of the Handshake Problem. Nearly everyone in class made an important and unique contribution. I then showed the class a few other problems whose solutions paralleled our problem, but I again emphasized that what was important to remember from today was not the problem. **What was important to remember was the variety of problem-solving strategies that we used to solve the Handshake Problem.** There were lots of smiles as the students gathered their notebooks and backpacks to leave class.

Look Back at the Handshake Problem

Recall Polya's four steps for solving problems. We will now do Step #4 and look back at the Handshake Problem so that if we ever see it or one like it again we will have strategies to try. This is where we find the "grain of discovery" that mathematician Polya said is "in the solution of any problem." This is how we gain confidence and build our bag of tricks to solve other problems. Notice that the class spent most of the time understanding the problem (Step #1) and devising plans (Step #2) to find the number of handshakes for 100 participants in the workshop. Toward the very end, Maria's pattern presented a way of using the number of participants to find the number of handshakes. Only then was the class able to "carry out the plan" (Step #3) and multiply "100 times 99 divided by 2" to solve the problem. The class spent the rest of the time looking back (Step #4) at initial guesses and a diagram made by two students. Even though they initially jumped right into "solving the problem" by making guesses, they still had to make great use of Polya's Steps #1 and #2 in order to check the guesses, which did contain important numbers but those numbers were not the correct answer to the question asked.

Now you and I are again using Step #4 to solidify what was done in our minds for future use. Incidentally, if you read Story #3 in this chapter about Gauss, you may have a thought about Elizabeth's recognition that 100 participants would make $1 + 2 + 3 + \ldots + 97 + 98 + 99$ handshakes. You might have thought to use Gauss's idea of pairing the first and last numbers and so on.

$$1 + 99 = 100$$
$$2 + 98 = 100$$
$$3 + 97 = 100$$
$$\ldots$$
$$47 + 53 = 100$$
$$48 + 52 = 100$$
$$49 + 51 = 100$$

This would be 49 pairs that add to 100. The 49 pairs include every number from 1 up to 99 *except* 50. So Elizabeth's method shows that there would be $1 + 2 + 3 + \ldots + 97 + 98 + 99 = 49 \cdot 100 + 50 = 4{,}950$ handshakes—the number of handshakes that my students found in class!

Before reading on, look back at the list of problem-solving strategies that I called "Your Bag of Tricks to Try." Check those that you believe my students and I used in the handshake problem story. Then compare your list with my list.

"Tricks" or Strategies Used by My Class to Solve the Handshake Problem	How the Strategies Were Used
State the goal.	The goal was the total number of handshakes, not time.
Throw out irrelevant information.	Don't use the number of men and women, political parties, or offices.
Try something "off the wall."	Maria's multiplication pattern was "off the wall."
Guess and check.	Raul and Teresa guessed. The rest of our work was checking. Their guesses involved important numbers, but they were not correct answers. Guesses seldom are. They are only guesses.
Make a picture or diagram.	Jorge and Donna made a diagram using dots for people and lines for handshakes.
Organize the information.	The t-table organized the data.
Identify patterns.	Maria, Bob, and Elizabeth found patterns in the t-table. Patterns solved the problem.
Generalize.	The class concluded that the number of handshakes is the number of participants times one less than the number of participants divided by 2.
Confirm the answer.	Jorge and Donna helped us review with their diagram. We discussed the initial guesses.

I will now tell you, the reader, what I told my students to remember. **It is the problem-solving strategies that are important to remember from this chapter, not the problems!** I would recommend that you now look again at Stories #5–8 to see which of the problem-solving "tricks" you might practice. Pick just one problem because I think that you know by reading about the experience of my class that solving problems takes time and willingness to try different strategies. Also, you may have noticed that one problem can be solved different ways and that exploring those ways gives us valuable experience, confidence, and better recall. Stories #5–8 are worked out for you in the Appendix.

OTHER STRATEGIES TO HELP YOU SOLVE PROBLEMS

1. Choose a work location where you can concentrate.
2. Keep working on the problem but break often and let the problem go for the moment. Persevere.
3. Ask questions about the problem.
4. Read the problem aloud more than once. Read the problem backward word for word to catch ideas you have previously ignored.
5. Review previous sections of your textbook in case you have missed something that would help with this problem.

ACT FOR SUCCESS | CHAPTER 9

1. Copy Polya's four steps. Notice how often you can use them in everyday life's problems.

2. Copy the list of "Tricks to Try." Choose one of Stories #5–8 from this chapter and write which "tricks" or strategies might be used on that problem.

3. Choose a math story problem from your textbook. Copy the problem and write how you would solve it. Identify the strategies from "Tricks to Try" you have used and how you used them.

4. Which of the "Tricks to Try" are basic steps necessary to solve any math story problem? Explain your choices.

MASTER MATH'S MYSTERIES

Dividing Fractions

Get Ready

To understand dividing fractions, first consider dividing whole numbers. "Six divided by two" can be thought of as "How many twos are there in six?" "One hundred divided by ten" can be thought of as "How many tens are there in one hundred?" We will think about fraction division in this way.

We will also use the Egg Carton Calculator introduced in Chapter 5. Remember:

$\frac{1}{12}$ is one pocket.

$\frac{1}{6}$ is two pockets, or $\frac{2}{12}$.

$\frac{1}{4}$ is three pockets, or $\frac{3}{12}$.

$\frac{1}{3}$ is four pockets, or $\frac{4}{12}$.

$\frac{1}{2}$ is six pockets, or $\frac{6}{12}$.

/////////	/////////	/////////			
/////////	/////////	/////////			

Divide These Fractions

a. $\frac{1}{3} \div \frac{1}{12}$ Think of this problem as "How many one twelfths are there in one third?" Using the egg carton, recall that $\frac{1}{3}$ is four pockets, or $\frac{4}{12}$. Also recall that $\frac{1}{12}$ is one pocket. Then rephrase the problem, "How many one twelfths, or pockets, are there in one third, or four pockets?" The answer is 4.

/////////	/////////				
/////////	/////////				

Conclusion: $\frac{1}{3}$ divided by $\frac{1}{12}$ is 4.

b. $\frac{2}{3} \div \frac{1}{12}$ Think, "How many one twelfths are there in two thirds?" Looking at the egg carton split into three equal pieces, notice that $\frac{2}{3}$ has eight pockets, or eight twelfths. The problem "$\frac{2}{3}$ divided by $\frac{1}{12}$" becomes "How many one twelfths, or pockets, are there in two thirds, or eight pockets?" The answer is 8.

/////////	/////////	/////////	/////////		
/////////	/////////	/////////	/////////		

Conclusion: $\frac{2}{3}$ divided by $\frac{1}{12}$ is 8.

c. $\frac{2}{3} \div \frac{1}{6}$ Think, "How many one sixths are there in two thirds?" Recall that $\frac{2}{3}$ is eight pockets of the egg carton. Also recall that $\frac{1}{6}$ is two pockets. This division problem then becomes "How many twos are there in eight?" The answer is 4.

/////////	\\\\\\\\\	/////////	\\\\\\\\\		
/////////	\\\\\\\\\	/////////	\\\\\\\\\		

Conclusion: $\frac{2}{3}$ divided by $\frac{1}{6}$ is 4.

(continued)

d. $\frac{3}{4} \div \frac{3}{4}$ Think, "How many three fourths are there in three fourths?" The answer has to be one.

/////////	///////	///////	\\\\\\\	\\\\\\\	\\\\\\\
\\\\\\\	\\\\\\\	\\\\\\\			

Conclusion: $\frac{3}{4}$ divided by $\frac{3}{4}$ is 1.

Summarize

The preceding examples can be generalized into a procedure.

a. $\frac{1}{3} \div \frac{1}{12}$ **is 4.** Notice that by "flipping over" the fraction on the right, $\frac{1}{12}$, to get $\frac{12}{1}$ and then multiplying $\frac{1}{3} \cdot \frac{12}{1}$ right straight across, the top times the top and the bottom times the bottom, we get:

$$\frac{1}{3} \div \frac{1}{12} = \frac{1}{3} \cdot \frac{12}{1} = \frac{12}{3} = 4.$$

b. $\frac{2}{3} \div \frac{1}{12}$ **is 8.** "Flipping over" $\frac{1}{12}$ (on the right) and then multiplying straight across gives the answer:

$$\frac{2}{3} \div \frac{1}{12} = \frac{2}{3} \cdot \frac{12}{1} = \frac{24}{3} = 8.$$

c. $\frac{2}{3} \div \frac{1}{6}$ **is 4.** "Flipping over" $\frac{1}{6}$ (on the right) and then multiplying straight across gives the answer:

$$\frac{2}{3} \div \frac{1}{6} = \frac{2}{3} \cdot \frac{6}{1} = \frac{12}{3} = 4.$$

d. $\frac{3}{4} \div \frac{3}{4}$ **is 1.** "Flipping over" $\frac{3}{4}$ (on the right) and then multiplying straight across gives the answer.

$$\frac{3}{4} \div \frac{3}{4} = \frac{3}{4} \cdot \frac{4}{3} = \frac{12}{12} = 1.$$

Go Back to Class

In math class, we use the words "take the reciprocal" instead of "flip over." So, in class, you will often hear, "To divide fractions, take the reciprocal of the fraction on the right and multiply." This is often abbreviated as "Multiply by the reciprocal." My only problem with this abbreviation is that students forget when to "multiply by the reciprocal" and they forget that they have to take the reciprocal of the right fraction, not the left fraction. Some students mistakenly use this procedure on multiplication problems, and some students "flip" the wrong fraction.

The Right-Hand Motion Divides Fractions

I believe (especially for you who are auditory learners or who have verbal-linguistic intelligence) that if you are going to verbalize a math rule in your mind, it is best to learn it with all of the details. My preference is to learn this sentence: "To divide fractions, flip the guy on the right and multiply straight across." To cement this procedure in your mind, especially if you are a kinesthetic learner or have Bodily-Kinesthetic Intelligence, repeat this sentence as you perform the following actions: Flip your *right* hand over, then hold two fingers straight out as you move your hand to the right. The two fingers remind you to multiply straight across–top times top and bottom times bottom. The actions accompanied by the detailed sentence spoken aloud will cement the procedure for dividing fractions into your mind. If you do the actions and say the sentence as you write out problems, the procedure becomes yours for life. You will eventually be able to do weird-looking algebra division problems without flinching.

Try These Examples

Check with the solutions in the Appendix.

1. $\dfrac{2}{3} \div \dfrac{2}{3}$ 2. $\dfrac{3}{4} \div \dfrac{1}{4}$ 3. $\dfrac{1}{4} \div \dfrac{1}{12}$

4. $\dfrac{5}{6} \div \dfrac{1}{12}$ 5. $\dfrac{1}{2} \div \dfrac{1}{6}$ 6. $\dfrac{4}{5} \div \dfrac{2}{10}$

Dividing Signed Numbers

The secret to dividing with signed numbers is to recall that division is related to multiplication in this way:

$$20 \div 4 = 5 \text{ because } 5 \cdot 4 = 20$$

We consider all the possibilities for dividing using signed numbers here.

Case 1: Positive divided by Positive = _____
Consider $8 \div 2$. The answer must be positive 4 because $4 \cdot 2 = 8$. So **a positive divided by a positive is positive** in the same way we already know.

Case 2: Positive divided by Negative = _____
Consider $8 \div (-2)$. The answer must be negative 4 because $-4 \cdot (-2) = 8$ as we discovered in Chapter 8's "Master Math's Mysteries." So **a positive divided by a negative is negative.**

Case 3: Negative divided by Positive = _____
Consider $-8 \div (2)$. The answer must be negative 4 because $-4 \cdot (2) = -8$ according to our work in Chapter 8's "Master Math's Mysteries." So **a negative divided by a positive is negative.**

Case 4: Negative divided by Negative = _____
Consider $-8 \div (-2)$. The answer must be 4 because $4 \cdot (-2) = -8$ according to our work in Chapter 8's "Master Math's Mysteries." So **a negative divided by a negative is positive.**

(*continued*)

♫ A memory technique for recalling the four cases for dividing with signed numbers is to sing the following lyrics to the tune of the song "If You're Happy and You Know It, Clap Your Hands." ♪

Dividing Signed Numbers ♪

A minus by a minus is a plus.   [Case 4]

A plus by a plus is a plus.   [Case 1]

A minus by a plus   [Case 3]

Or a plus by a minus   [Case 2]

Is a minus, yes, a minus,

Not a plus.

Try these:

7. $-80 \div (-10)$ **8.** $(-55)/5$ **9.** $9 \div (-3)$

10. $(-90) \div 3$ **11.** $-125/(-25)$ **12.** $10 \div (-\frac{1}{2})$

Look Back–Comprehensive Review for Your Practice

Perform the indicated operation:

1. $0.53 + 2.132$ **2.** $-5(-4)$ **3.** $3 - (-7)$

4. $\frac{1}{7} \cdot \frac{9}{5}$ **5.** $\frac{1}{6} + \frac{1}{4}$ **6.** $\frac{2}{3} - \frac{1}{12}$

Try these order of operation problems:

7. $9(2) - 4(6 \div 3)^2 + 6$ **8.** $13 + 3(5 - 4)$

Perform the indicated operation:

9. $2,770 + 230$ **10.** $9.4 - 4.2$

Put it together. Practice vocabulary and fractions.

11. Find the sum of $\frac{4}{15}$ and $\frac{9}{15}$.

12. Find the quotient of $\frac{2}{3}$ and $\frac{1}{3}$.

13. Find the product of $\frac{3}{8}$ and $\frac{1}{3}$.

14. Find the difference of $\frac{7}{12}$ and $\frac{2}{12}$.

Student Success Story

Advice for Struggling Math Students

"This sounds nuts—my best advice to struggling math students is to quit struggling. I wish I knew then what I know now—which is to relax, just in all areas. When I relax, I find that I'm much more available to get it, to learn it. But I was so convinced that I couldn't do it. While I was being taught, I would spend time in my own mind telling myself that I couldn't do it. I was so uptight about not being able to do it that I didn't know how to stop struggling against it. And given the chance to struggle a little, I struggled a lot. The best thing that I could have done would be to just quit struggling against it. And let the grades go."

Giovanna Piazza, Priest and ethics instructor

"Take all the barriers you have about math—all the preconceived notions that you have about math as a subject—put them in a little box and burn it or bury it. Then relax with math and let it infuse you instead of worrying about what you're going to do with it. Just let it come in to you. [L]ook at it and notice the relationships and places where you will use the concept. You can make up a formula once you have a concept. I didn't know that. Once you have the pattern then you can do not only math but many things."

Bobbi Nesheim, Ph.D.,
psychotherapist and owner of
Center for Creative Growth

## CHAPTER 10

Manage Math Anxiety

"I've been absolutely terrified every moment of my life—and I've never let it keep me from doing a single thing I wanted to do."

GEORGIA O'KEEFFE

Taking Charge

For a useful perspective on anxiety, read the following sentence and consider the underlined words.

> Anxiety results when you are <u>required</u>
> to stay in an uncomfortable situation
> where you <u>believe</u> you have no <u>control</u>.

To lower your math anxiety, consider the three underlined words.

1. Reframe <u>Requirements</u> into Choices

As a child, you depended upon the adults in your life and were <u>required</u> to participate in many experiences—in particular, math experiences—whether you liked them or not. Fortunately, as an adult, you have more choices. Although you cannot control everything happening in your life, if you choose to do so, there is much you can <u>control</u>.

As you choose a major or vocation with a math component, you place yourself voluntarily in a math class. You are now choosing the math requirement. Every single time you attend class, remind yourself that you have chosen this hard work to gain something you want—the profession of your choice.

> A student once told me that her breakthrough in math came when she realized she wanted to be a teacher **more** than she was afraid of math.

2. Identify Your Math <u>Beliefs</u> and Reframe the Negative Ones

Your beliefs affect your actions. If you **believe** you cannot do math, you won't experiment with it or practice it or place yourself where you can learn more. If you *believe* people who are "good" in math do math quickly, you will be impatient with yourself when you learn. You won't allow yourself the necessary "percolating" time for math to settle into your mind. Chapter 1 identifies your math beliefs with a true/false quiz and gives you alternative reframes of negative beliefs.

3. Choose to Take <u>Control</u> of All Your Math Activities

You have assumed **control** now by reading this book, by answering the questions, by examining your previous math habits, by experimenting with math problems at your level, by recognizing your right to base your self-esteem on things other than your math skills, and by asking questions when you don't understand. This book is filled with activities and short-term goals you can adopt that give you control of math and help you create flow with math. *As you examine your math beliefs in this book and change requirements to choices, you take charge of your math life and your anxiety level will decrease.*

Stress responses occur as overwhelming negative feelings and debilitating physical symptoms. Stanford neuroscientist Robert Sapolsky's research (2000) shows that "humans can activate stress responses by thoughts." Changing your thoughts can de-activate your stress responses. This chapter and Chapter 11 will teach you tools to change your thoughts, beliefs, and also behaviors.

Thoughts Are in Charge

Almost everything we experience as human beings can be categorized as a thought, emotion, behavior, or body sensation. These four categories are distinctly different, yet they are interrelated. Each one influences and is influenced by the others, as shown in Figures 10.1 and 10.2 (Greenberger & Padesky, 1995).

- **Thoughts** include your beliefs, ideas, opinions, knowledge, rules, thinking, and experiences.
- **Emotions** are feelings and moods.
- **Behaviors** are how you act—what you do.
- **Body sensations** are the physical reactions of your body.

Compare the two contrasting interrelationship charts shown in Figures 10.1 and 10.2. See how the four aspects of our experience cause one another. Specifically, start by reading the "thoughts." Notice that changing the thought changes the other three components.

It is through your thoughts and behaviors that you can consciously intervene in these interrelationships to change the other outcomes. That means that changing your thoughts and behaviors with math can neutralize negative emotions and body sensations. See the contrast from Figure 10.1 to Figure 10.2 by just changing the thought from "I can't do math" to "I can do some math."

Figure 10.1   *The Effects of a Negative Thought*

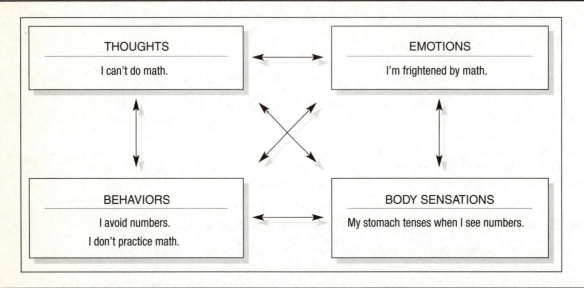

If you say "I can't do math," you will likely avoid numbers and not practice. Then you may become frightened by math and your stomach may tense. But you could reframe your thought "I can't do math" by saying, "I can do some math. I can learn more. I don't need to get it all right now." Then you could take a deep breath, write the problem and a possible solution, or get help. You might feel relieved and perhaps curious. You may even experience some joy at the math skills you do have. Your body may relax as you become calm and your heart rate slows.

Figure 10.2   *The Effects of Alternative Thoughts*

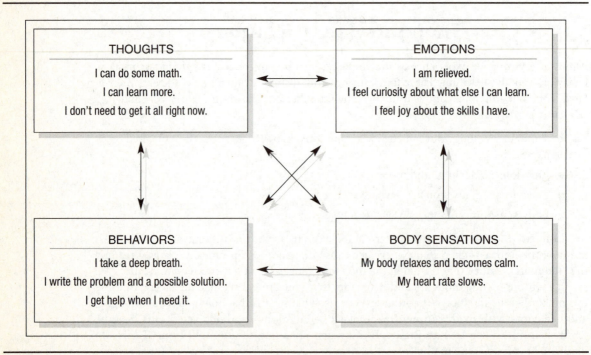

Reframe the Mean Math Blues

Figure 10.3 lists the symptoms of math anxiety, or what I call the "Mean Math Blues." Notice how these negative thoughts, behaviors, emotions, and body sensations are all related to one another. Each one causes and is caused by the others. Circle any thoughts that you have frequently and would like to change. The act of recognizing your negative thoughts and behaviors begins the process of change. Your control lies in changing your thoughts and behaviors about math. Changing those thoughts and behaviors will begin to alter the emotions and body sensations that keep you uncomfortable.

Every one of the negative thoughts in Figure 10.3 can be reframed in a neutral or more positive way. A reframe considers the same situation but brings a new perspective to it. Reframing is an effective way to change our thinking and then our attitudes. The Introduction of this book describes reframing in more detail, and Chapter 11 gives a step-by-step way to reframe negative thoughts and to choose useful interventions.

Figure 10.3 *Symptoms of the Mean Math Blues*

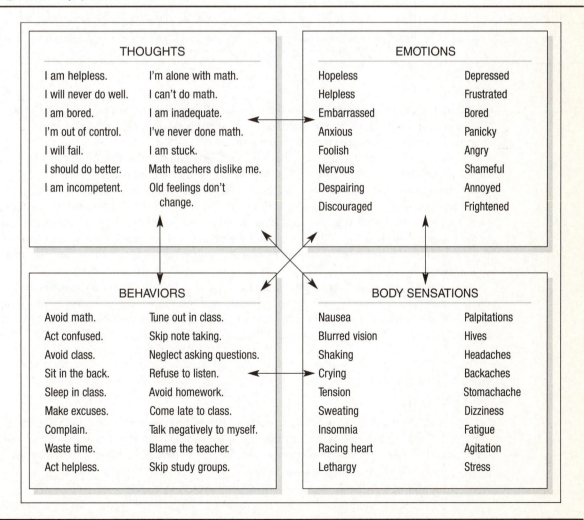

THOUGHTS		EMOTIONS	
I am helpless.	I'm alone with math.	Hopeless	Depressed
I will never do well.	I can't do math.	Helpless	Frustrated
I am bored.	I am inadequate.	Embarrassed	Bored
I'm out of control.	I've never done math.	Anxious	Panicky
I will fail.	I am stuck.	Foolish	Angry
I should do better.	Math teachers dislike me.	Nervous	Shameful
I am incompetent.	Old feelings don't change.	Despairing	Annoyed
		Discouraged	Frightened

BEHAVIORS		BODY SENSATIONS	
Avoid math.	Tune out in class.	Nausea	Palpitations
Act confused.	Skip note taking.	Blurred vision	Hives
Avoid class.	Neglect asking questions.	Shaking	Headaches
Sit in the back.	Refuse to listen.	Crying	Backaches
Sleep in class.	Avoid homework.	Tension	Stomachache
Make excuses.	Come late to class.	Sweating	Dizziness
Complain.	Talk negatively to myself.	Insomnia	Fatigue
Waste time.	Blame the teacher.	Racing heart	Agitation
Act helpless.	Skip study groups.	Lethargy	Stress

Negative Thoughts About Math	Possible Reframes of Each Negative Thought
I am helpless.	I feel helpless at this moment.
I'm alone with math.	I feel alone right now. I do have resources and choose to use them.
I will never do well.	I can't predict the future. My determination will help me learn and accomplish what I wish.
I can't do math.	I can do some math and I can learn more. (See Chapter 1 and Figure 10.2.)
I am bored.	I am bored with math at this moment. I can do my math work anyway. Being engaged with the math may or may not change my boredom but at least I will complete my course.
I am inadequate.	I feel inadequate right now. I can get support.
I'm out of control.	I feel out of control right now. Taking charge of my behavior can change that feeling.
I've never done math.	I know some math. I can learn more.
I will fail.	I can't predict the future. I can take charge right now and make changes.
I am stuck.	I feel stuck for the moment. I can back up to something I understand and call in my resources to assist me in moving forward.
I should do better.	I want to improve my math skills and will now take more control in using my time and resources.
Math teachers dislike me.	I may have had a negative experience in the past with one math teacher. Math teachers are all different. I will seek a patient, clear, and encouraging teacher to learn math differently than in my past.
I am incompetent.	I feel incompetent for the moment. I am competent to get assistance when I need it.
Old feelings don't change.	Feelings are fluid and depend on my thoughts and behaviors. I am reframing my negative math thoughts and choosing different proactive behaviors.

Reframes of the negative thoughts from Figure 10.3 appear in the nearby list. Reframing your negative math thoughts and changing your math behaviors changes your feelings and your body sensations related to math. See Figure 10.4.

As we consider our thoughts, behaviors, feelings, and body sensations, we recognize that it is very difficult, if not impossible, to change feelings and body sensations. We cannot expect immediate results when we say, "I am now joyful or calm or relaxed" or "The tightness in my shoulders or my headache will now go away." However, we can begin to change our thoughts and our behaviors so that those feelings and body sensations change. It is with your thoughts and behaviors that you have the power to intervene when you experience negativity in your math work. Chapter 11 will give you more tools to change negative math thoughts.

Practice reframing math thoughts. Read the following "Math Reframes" *aloud* to create a mental math picture that is supportive, hopeful, and strengthening. Coupled with asking questions and working math problems, these reframes open your subconscious mind to confidence with math and to positive feelings about math, if you repeat them often.

Math Reframes

I DO MATH.

- My mind is open and I invite math in. It is not a problem when I don't understand right away. With continued practice and many questions, my understanding increases. My goal is to progress—to be able to do a little more, step by step, each day. Understanding follows. It is O.K. to say, "I do not *yet* understand." I am willing to learn. I see the processes more clearly every day.

- I look back to notice how far I've come and I enjoy my progress. I know more than last week and much more than a month ago. Small steps add up to BIG changes.

Figure 10.4   *Effective Thoughts to Manage the Mean Math Blues*

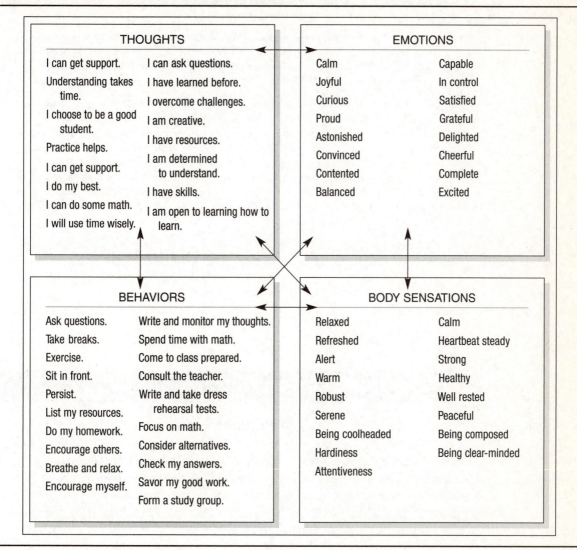

I <u>DO</u> MATH.

- My questions are in process of being answered. I continue to put them out there and answers come. It is O.K. to ask several people as many times as I need. When I walk into my math classroom, my mind opens wider to math. I take in positive suggestions only. Math class feels safer as I take charge of my learning.
- Working together with others helps. When I laugh, my mind opens to understanding, so I often look for humor in math. Each day I progress toward my math goals.

I DO <u>MATH</u>.

- My new understandings of math become stronger. The patterns are there and my eyes spot them more quickly each day. The more I practice, the more connections form in my brain. Understanding happens gradually. I release my old misconceptions about math and become curious about solving math mysteries.

<u>I DO MATH</u>.

ACT FOR SUCCESS | CHAPTER 10

1. Write down automatic negative thoughts that you experience with math work. Consider possible reframes for those thoughts. Show other people some of the reframes from this chapter, and ask for their assistance in reframing your personal negative math thoughts.

2. Read the Math Reframes at the end of this chapter aloud three or four times a week—more often at first. Copy them and post them where you will see them as you work. Keep a copy in your math book for review. (Change them, if you wish, to make them more effective for you. However, make certain that you keep them simple, positive, and in the present tense.) You may feel silly reading these reframes aloud at first. Remind yourself that this is an effective method to consciously influence your unconscious belief system about math. If feeling silly is the worst thing that can happen, what do you have to lose? The gains could be enormous.

MASTER MATH'S MYSTERIES

The Number Line

Imagine that you are going to ride on a subway train. You go to the subway station. Some trains travel in one direction and some in the opposite direction. How could you use just numbers to describe the difference between traveling 4 miles east and traveling 4 miles west? In Chapter 7's "Master Math's Mysteries," we said that all numbers are signed numbers and that the signs, positive and negative, represent direction. We could say that traveling 4 miles east on a subway train is represented by positive 4 and traveling 4 miles west is represented by negative 4.

In math we use a picture that shows positive and negative numbers on something that is a bit like a train track. We call that picture the "number line." From our original train station, the trains go east or go west. We call the original train station the zero position on the number line. The zero position is where the numbers change from negative to positive or positive to negative. When we draw our number line, we locate the zero position (our subway station) first. Then we mark the miles to the east (or right, with positive numbers) and to the west (or left, with negative numbers). Imagine traveling the following number line. Our beginning station is at the zero point. Where would we be if we traveled on the train 4 miles east? We would be at positive 4, shown by a dot on the number line at 4.

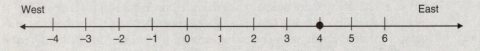

Where would we be if we traveled 3 miles west? We would be at negative 3, shown by a dot on the number line at −3.

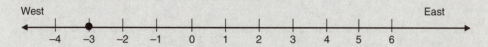

When we put the dot on the 4 and on the −3 of the number lines above, we say that we are graphing numbers on the number line. Graph the following positions on the following number line. Answers are in the Appendix.

1. 0 **2.** −2 **3.** 1

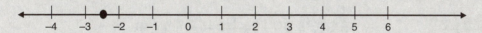

One major difference between our number line and the subway train track is that the number line is infinitely long in both directions. A second difference is that we can stop anywhere on the number line because we're not on a train. Suppose that we wanted to go 2.5 miles west. Where would that be located on the number line? We would put a dot halfway between −2 and −3:

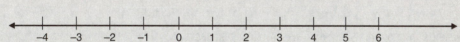

Graph these positions on the following number line.

4. 1.5 **5.** −3.5 **6.** 0.5

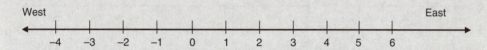

Now suppose that we wanted to go $2\frac{1}{4}$ miles east. Where would that be located on the number line? Previously, to find 2.5 miles west, we just found the halfway point between −2 and −3. Now to find $2\frac{1}{4}$, we need to do something different. We will go 2 miles east and then $\frac{1}{4}$ of the way between 2 miles and 3 miles farther. To visualize this, we need to split the number line between 2 and 3 into four equal pieces and travel only 1 of them.

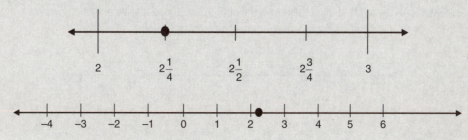

If we were to graph $2\frac{3}{4}$, we would still split the number line between 2 and 3 into four equal pieces but now we would travel 3 of them.

(continued)

Graph the following positions on the number line.

7. $-1\frac{1}{4}$ **8.** $-3\frac{3}{4}$ **9.** $4\frac{3}{4}$

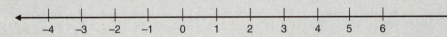

In the same way, if we graphed $2\frac{1}{6}$, we would split the number line between 2 and 3 into 6 equal pieces and travel 1. If we graphed -2.3, because -2.3 is $-2\frac{3}{10}$, we would split the number line between -2 and -3 into 10 equal pieces and travel 3 of them from -2 left (or west). Remember that we are traveling west (left) when we use negative numbers:

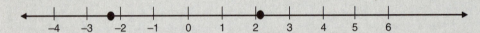

Graph the following positions on the number line.

10. $-3\frac{1}{8}$ **11.** 5.2   **12.** 0.7

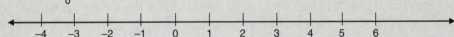

Angels on the Head of a Pin

Has anyone ever asked you a really deep math question? Here's one. Give it a try. Don't worry if you need some help to figure it out.

How many numbers are there between 1 and 2?

We know from our previous work that fractions and decimals will fall between the two numbers. One way to look at this question is by using a number line.

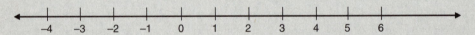

Remember that we place the zero first and then count off equal-sized spaces to the right and label them with the counting numbers. Next we start at zero again and count off the same-sized spaces to the left and label them with the opposite values, negative numbers.

To answer our question I'm going to enlarge the section around the 1 and 2.

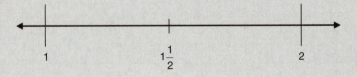

We can see that we can cut the space between the two numbers in half and even cut the two smaller sections in half again. We see four equally sized spaces between 1 and 2; that means we have fourths.

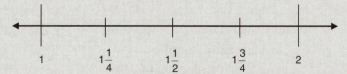

Looking at the number lines, we can see that $1\frac{1}{4}$, $1\frac{1}{2}$, and $1\frac{3}{4}$ are numbers between 1 and 2. But what if we cut each of the fourths again in half? We could because $\frac{1}{2}$ of $\frac{1}{4}$ is $\frac{1}{8}$. That would give us eight equal sections *and* several more numbers between 1 and 2. The number line would look like:

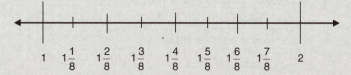

Now you might begin to get an idea of the answer to our question "How many numbers are there between 1 and 2?" We could continue the process indefinitely cutting our eighths in half, then sixteenths in half, and so on. We would soon run out of room to write the numbers under the line marks, but there would continue to be more numbers in between the others. Realize too that we have not even looked at thirds or fifths or many other denominators. You are probably beginning to see that this process could go on indefinitely.

So, how many numbers are there between 1 and 2? An infinite number.

Below is a picture of all those numbers between 1 and 2. The amazing part is that we cannot write them all, but we can make a picture that symbollizes all of them. And we call that picture a "graph."

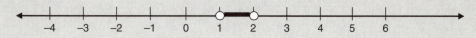

Student Success Story

Awakening

Though darkness seems warm comfort,
    hatchling knows the time
    to peck away an opening—
    long labor—then she's free.

Or off from seed, once roots reach deep,
    one stem must find its way,
    its steady climb past stony obstacles—
    long labor—though its destiny, it knows,
    is up and out, toward the sun.

And even pain can serve:
    in her own shell, what she cannot remove
    she builds a shield around:
    long labor—but the product of her industry—
    a pearl.

Victoria Stephenson,
English professor,
Santa Ana College

Reprinted with permission.

Neutralize Negative Math Thoughts and Behaviors

"All things change—so can we."

JULIA CAMERON

The previous chapter made the case for you to examine and change your thoughts and behaviors in order to lower your math anxiety and feel better with math. You learned about the power of reframing negative thoughts. This chapter will give you a step-by-step method for examining your thoughts and reframing them.

Thought Distortions in Automatic Negative Thoughts

Students experiencing math anxiety or the Mean Math Blues report negative thoughts such as "I can't do math" or "I will never pass" or "I'm the only one who doesn't understand." When thoughts such as these run through our minds over and over, they are "automatic negative thoughts" that influence us in a negative way.

These repetitive negative thoughts contain distortions that warp our thinking (Beck & Emory, 1985). We will discuss six Thought Distortions (Burns, 1999) in negative thinking that relate to math. As you recognize negative math thoughts, identify Thought Distortions, and use Intervention Strategies (Burns) from this chapter, you can consciously neutralize the negativity.

Figure 11.1   *Thought Distortions Found in Automatic Negative Thoughts*

THOUGHT DISTORTIONS
in Negative Thinking

See Figure 11.1 for the six Thought Distortions found commonly in negative math thinking. Then read through the definitions and examples for each. Afterward, you will be given exercises to help you practice identifying Thought Distortions and neutralizing negative thoughts.

1. Crystal Ball Thinking

Crystal Ball Thinking predicts the future and colors it negatively. Predicting the future ignores the fact that you change every day and moment of your life. Your life is a dynamic process of change. You are not a static being. Unfortunately, you can actually make bad things occur in your life by believing that the worst will happen and then behaving as if this were true.

Examples of Crystal Ball Thinking with Math

❒ I *will* never be able to do math.

❒ I *will* never graduate.

❒ If I don't understand a problem quickly, I never *will*.

❒ My memory of my negative math experience *will* never go away.

❒ I won't pass that math test next week.

2. Mind Reading

With *Mind Reading,* you imagine that you know what someone else is thinking or feeling. This causes you to draw inaccurate conclusions and behave based on a fantasy.

Examples of Mind Reading About Math

❒ I *think* the teacher doesn't want to help me, so I won't ask.

❒ The others in the class *think* I'm a math failure. They wouldn't like me to ask them a question, so I won't.

❒ Other students *think* I'm stupid when I ask questions, so I will stop.

❒ Mathematicians always do problems quickly in their heads. I should understand these math problems right away.

❒ I better not ask questions because the teacher and students will believe that I'm inadequate.

3. Overgeneralizing

Overgeneralizing makes broad conclusions, often dire predictions, about the future based on one event. When statisticians make generalizations, they ask *many* people or view *many* situations first. When you make sweeping life decisions based on one incident, you are overreacting.

Examples of Overgeneralizing About Math

❒ My *whole* life is ruined because of this problem.

❒ If I can't complete this homework, I will *never* be able to get math.

❒ Because I failed the test, I know I will *always* do poorly in math.

❒ I can't do math.

❒ I've used a math tutor. He was critical and unhelpful. I don't work well with math tutors.

❒ My high school math teacher was awful. Math teachers are insensitive.

4. Absolute Thinking

Absolute Thinking recognizes only two opposite extreme alternatives—*black or white*—*all or nothing.* According to this thinking, on a scale from 1 to 10, everything is either a 1 or a 10. Black-or-white thinking ignores the alternatives—the gray, in-between areas. Very few issues are *either/or.* Watch for **1 or 10, black or white, good or bad, all or nothing.**

Often we filter out and minimize the *positive* input while focusing on the negative. Obsessing on the few negative things that happen while ignoring the rest is a common human reaction. However, it is an inaccurate perception of the world. Amplifying the negative is so overwhelming that most people deny and avoid rather than using the negative information as feedback for learning and making changes.

Examples of Absolute Thinking with Math

❏ Either I get an A grade or I am a failure. Bs and Cs are unacceptable.

❏ Either I get these three math problems or I can't finish and understand my assignment.

❏ Math is always hard.

❏ Only smart people can do math.

❏ If I don't do well on this homework, I won't pass my class, finish my degree, or ever get a job.

❏ I look at my 75% test grade and only notice what I got wrong, not the fact that I got many things correct.

❏ I say something I regret in class and keep berating myself for not being perfect.

❏ I got an A on my first three exams. Now I got a C. I knew I couldn't do math.

❏ I cannot do a math problem, so I think I cannot do the whole homework assignment and probably cannot do any math from here on.

❏ I work with a study group and am uncomfortable about one incident during the four-hour meeting. The next day I have maximized this negativity to the entire session and vow to never attend one again.

5. Feeling = Being

Feeling = Being equates the *feelings* you have about yourself or your moods with who you truly are as a person. You forget that you are not your feelings. Your feelings are the results of your thoughts and behaviors and are separate. Your feelings are part of you, but they are not accurate pictures of your whole self. **The truth: Feeling is *not the same as* being.**

Examples of Feeling = Being with Math

❏ I feel anxious about math or math tests, so I think I am unable to do well in math or will perform poorly on tests.

❏ Because I feel inadequate, I believe I am inadequate.

❏ I feel shy, so I think I am too shy to ask questions. (See Chapter 1.)

❏ I feel overwhelmed, so I think I can't learn this.

❏ I feel bad or dumb or guilty, so I think I am bad, dumb, or guilty.

6. Shoulding

Shoulding uses words like "should," "must," "have to," and "ought." These words imply a negative, parent-like order. Most people don't accept orders well. These words attempt to place blame. You may feel coerced and perhaps a bit rebellious when spoken to in this manner—whether by yourself or someone else. Your choices and your goodwill shrink under orders. **Watch for: I should . . . , I must . . . , I have to . . . , I ought to. . . .**

Examples of Shoulding with Math

❏ I *should* do all my homework.

❏ I *shouldn't* make mistakes.

❏ I *must* get very good grades.

❏ I *ought* to be perfect.

❏ I *have* to be good.

❏ I *should* be the best.

❏ I *ought* to do this.

❏ I *shouldn't* count on my fingers.

To counteract Shoulding, you can make many thoughts more reasonable and truthful by reframing the previous statements with "**I choose to**" or "**I want to**" or "**I will.**"

- I choose to do all my homework.
- I will make mistakes. (It's human.)
- I want to get very good grades.
- I want to be perfect. (It's not human.)
- I want to be good.
- I would like to be the best.
- I choose to do this.
- I choose to use my fingers when I need to.

Practice Identifying Thought Distortions

For practice, analyze the following automatic negative thoughts and identify the thought distortions (Crystal Ball, Mind Reading, Overgeneralizing, Absolute Thinking, Feeling = Being, or Shoulding). You may recognize more distortions than I have.

1. I will fail my test next week.
2. I can't do math.
3. I should be getting a better grade.
4. The other students think my question was dumb.
5. The teacher doesn't want us to ask questions.
6. I will never understand word problems.
7. Word problems are always hard.
8. Tutors don't work well for me. I have had a bad experience.
9. I am hopeless with math.
10. I am stupid.

I identify the following thought distortions in the automatic negative thoughts just listed. You might see others.

1. Crystal Ball
2. Overgeneralizing or Absolute Thinking
3. Shoulding
4. Mind Reading
5. Mind Reading or Overgeneralizing
6. Crystal Ball or Absolute Thinking
7. Absolute Thinking or Overgeneralizing
8. Overgeneralizing
9. Feeling = Being
10. Feeling = Being

Editing Negative Math Thoughts

Now practice both identifying Thought Distortions and reframing negative math thoughts in two math situations.

Read Situation 1 and the automatic negative thoughts listed below it. Then identify the Thought Distortions and reframe the automatic negative thoughts.

Situation 1: Homework

You are doing well with your math homework but get stuck on several problems in a row. You might automatically think some of these negative thoughts:

a. "I will never understand math."

b. "I am stuck here forever and I will be unable to do the rest of the assignment."

c. "I feel dumb and stupid."

d. "I cannot do math."

e. "Math problems contain tricks meant to stump me. Math is out to get me."

Before reading on, identify the Thought Distortions contained in each of the negative thoughts listed. Write them down.

These are the Thought Distortions I identify. You might find more. As you read my answers, begin to reframe each of the negative thoughts by neutralizing it and removing the Thought Distortion.

a. **"I will never understand math."**   Thought Distortions could be Crystal Ball (because of the word "will") or Absolute Thinking (because of "never").

b. **"I am stuck here forever and I will be unable to do the rest of the assignment."** Thought Distortions are Crystal Ball (because of "will"), Absolute Thinking (because of "forever"), and Overgeneralizing (because of "stuck" and "forever").

c. **"I feel dumb and stupid."**   This is Feeling = Being. Feeling dumb or stupid does not mean that we are dumb or stupid. It only means that we are confused at this moment.

d. **"I cannot do math."**   This is Absolute Thinking or Overgeneralizing. In saying this, we assume all or nothing and we are likely taking one or two situations and making a major generalization.

e. **"Math problems contain tricks meant to stump me. Math is out to get me."** Distortions here are Overgeneralizing or Mind Reading. We assume nasty tricks in all math and we make judgments about the intentions of the authors of the problems.

Before reading on, reframe each negative thought into a more neutral and useful statement. Make it a true statement to be effective. Then read the following suggestions, keeping in mind that there are many, many ways to reframe:

a. "This has happened to me before and I have worked through it."

b. "Just because these few problems are difficult doesn't mean all the rest will be difficult too. This is an opportunity for me to figure out what I misunderstood and correct it. Maybe I need to ask some questions or to do some of the examples again."

c. "Feeling dumb or stupid doesn't mean that I am dumb or stupid. The learning process is challenging. There must be something I don't quite understand here. I can use the answer from the answer key and try to work backward to get the process."

d. "I know some math and can learn more. I have many resources to assist me—the book, notes, examples, the instructor, friends, and tutors."

e. "I could look ahead and see if there are some problems that I can do. I will mark these problems so I can come back to them."

Situation 2: Consulting the Teacher

You finally get the courage to go to the teacher's office to ask questions. The teacher looks at your math work and says, "You should know this."

You may feel embarrassed, but shaming may not be the intention of the teacher at all. Most teachers are patient and willing to work with you.

First, write down negative thoughts some students might automatically think in this situation. Second, identify the Thought Distortions in your examples and reframe each negative thought into a more neutral and useful statement. Make your reframe a true statement to be effective.

I might think:

a. "I should know much more than I do." This is Shoulding. Reframe: I would like to know more than I do and right now I don't. I could use some assistance or direction to other resources.

b. "Oh, wow, I'm in trouble now." This is Mind Reading. Reframe: It is true that there are important pieces of our math work that I don't know. Perhaps the teacher is reminding me that I need to study differently and spend more time or that I am in a class that is too high for my current skill level.

c. "I'm stupid." This thought distortion is Feeling = Being. Reframe: I may feel stupid, but I am not stupid because I am here to get the teacher's assistance and advice. My skill level may not match the challenges of this class. I need to figure out whether I need to focus and spend more time at this level or to retake an earlier level to build my skills.

d. "I will never understand math." This is Overgeneralizing or Absolute Thinking. Reframe: Not understanding the current concepts does not mean I will never understand. It means that my current skills don't match the challenges. This is not an all-or-nothing situation. I have learned before. The teacher or someone else can help me find the level where I work best. I need to examine my own study behaviors to make the changes that will further my understanding.

Here are more suggestions for reframing in this situation. I suggest that you read them aloud to place positive messages about asking questions into your mind. The first suggestion is my best idea of what you might say in response to your teacher's comment.

■ You might say to your teacher, "Yes, you're right. I should know this and I don't. Would you help me figure out how to best learn this and get my questions answered? I care about learning this material and passing this course. I appreciate your time and assistance. Do you have any further recommendations for places or ways to get assistance? This is what I am able to do now. Perhaps there are examples in the book you could recommend, or perhaps you could help me identify the error in my thinking that is causing me so much difficulty with this concept."

■ You might say, "I wish I knew this *and* I don't. I very much want to learn this. I need a resource that will help me to start improving my understanding."

■ You might say to yourself, "Whew. I have come to the right place to find out how to further my understanding and learn what I 'should' know in order to advance in this course. If I am in the wrong course, it is best to find out now and get into the right course to lower my frustration."

Intervention Behaviors to Neutralize Negativity

"People do not usually overcome anxiety until changes in thoughts are accompanied by changes in avoidance behaviors" (Greenberger & Padesky, 1995). This means that you will want to change what you do to avoid math. Here are eight powerful behavioral interventions (Burns, 1999) that could assist you in changing what you do and neutralizing your negativity about math. Check the ones that appeal to you now as possibilities. Some of these strategies might need to "grow on" you awhile to become acceptable. Return to this section and reread the possible intervention strategies from time to time. You may discover more strategies that will be useful.

1. Examine the Evidence

What is the evidence that your negative thought is true? Are you overreacting? What is the evidence that the thought is false? What would you do differently if this thought were false?

Example

You think you will fail math. Ask yourself the following questions.

- Have you truly gotten failing test scores?
- Are the low grades the result of your neglecting studying and homework?
- Have you refused to get assistance or ask questions?
- What is the evidence that you are, in fact, failing?

2. Get a Different Perspective

The following actions can help you get that new perspective:

- Tell yourself *what you would tell a close friend* who has this thought. Would you want him to be handicapped by this negative thought? We are often much harder on ourselves than we are on others. Could you not choose to be as kind to yourself as you would to your friends?
- Speak to yourself or write down *what a good friend would say to you* about this negative thought. A close friend is probably more objective and positive than you would be for yourself.
- You may feel like you are the only one struggling. Talk to your teacher, your tutor, and other students in the class to see how realistic your thought is.

3. Do Something Differently

Behave in a new way to get a different result. Identify actions that contribute to your negative math thoughts and learn from what you have recognized. Change these actions to behaviors that are new to you.

Examples

- Recognize that you cannot expect yourself to understand math when you do not practice by doing your homework.

- Recognize that you may not understand the class lectures because you don't ask enough questions or take enough notes or get enough sleep to stay alert.
- Reread the checklists in Chapters 4 and 5 for new ideas for being a successful student.

4. Track the Number

Record the number of times that you think these negative math thoughts. Recognition is one of the first keys to bringing them to consciousness and changing them.

5. Identify the Worst-Case Scenario

Ask yourself, "What is the worst thing that can happen in this situation?" Often the worst thing that can happen isn't so bad after all. You can survive all kinds of "terrible" things. Sometimes it is the fear that is worse than the consequence.

Example

You have a question in your math class and are reluctant to ask it. The worst thing that could happen might be that the teacher will yell or that the teacher will refuse to answer or that someone might laugh. None of those three events is life threatening. Each possibility shows insensitivity on someone else's part—not an indication that you are wrong for asking your question. Sometimes we are willing to risk possible consequences when we realize how unlikely they are to occur.

6. Change the Wording

Restate the thought in a way that is neutral or could actually be positive. Add the words "right now," "for now," or "yet." (This is reframing.)

Examples

- Change: "I will fail math" to "Right now I cannot predict the future and I can certainly do some things to prevent failure."
- Change: "I can't do math" to "Right now I am unable to do these math problems."
- Change: "I don't understand" to "I don't understand *yet.*"
- Change: "I'm not prepared" to "I'm not prepared *yet.*"
- Change: "I'm never going to get this" to "I get it up to this point."

7. Act "As If"

Act as if you had whatever trait you lack or are whatever you would like to be. Ask yourself, "How would I look? What would I hear differently? What would I say? How would I behave?" Assume new thoughts and behaviors—don't just pretend. "Try on" success.

Examples

- If you want to be a successful math student, consider how good math students act. What behaviors do they do? Most importantly, they never pretend that they understand something when they don't. They go to class, ask questions, complete homework, work with classmates and teachers, read the textbook, use tutoring services, admit they don't understand. . . . What else would a successful student do?

- Public speakers frequently act "as if" they feel confident. They put their shoulders back, smile, speak up, make eye contact, and act as if they know what they're talking about. As Anna sings in the musical *The King and I,* they fool themselves as well.

8. Affirm Your Best

Read the "Math Reframes" at the end of Chapter 10 aloud to create a mental math picture that is supportive, hopeful, and strengthening. Coupled with asking questions and working math problems, these statements open your subconscious mind to confidence with math and to positive feelings about math, if you repeat them often.

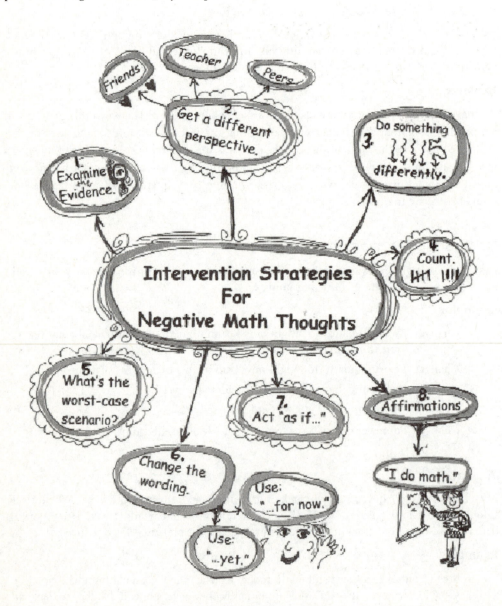

ACT FOR SUCCESS | CHAPTER 11

1. Copy the six Thought Distortions and list one example of each of them that is not listed in this book. For each example, write a reframe that eliminates the Thought Distortion.

2. List five negative math thoughts you have had yourself or heard others express. Identify the Thought Distortions. Write down alternative neutral thought substitutions for the negative thoughts.

3. Analyze this situation: You work really hard on a math problem and can't do it. When you go to class, you mention that you didn't get it. You hear another student say, "That was easy." List possible automatic negative thoughts and their Thought Distortions. Which of the following alternative related thoughts seem reasonable to you?

   ■ "Most things are easy once I understand and practice them."

   ■ "That other student could be expressing relief or enthusiasm that she understands or she might just be bragging and insensitive. I do not have to hear her comment as criticism of me. My responses are my responsibility. I can't change the world—only my reaction to it."

   ■ "Right now I don't understand this problem and I, being human, have a right to the learning process and its challenges. People do not all learn at the same rate and that's O.K. I am not in competition with the others in my class. Learning math is not a race."

4. Copy the eight Intervention Behaviors to Neutralize Negativity. Choose one of your bothersome negative thoughts about math that you wish to change. Write how one of the Intervention Behaviors could assist you in changing the negative thought.

MASTER MATH'S MYSTERIES

The Master Math's Mysteries sections of Chapters 1 through 10 have given you a review of basic math skills that are life skills. The next step in the math sequence would be prealgebra. Chapter 12 introduces you to algebra and gives you an overview framework for understanding it.

Working the problems here will give you a review of all of the Master Math's Mysteries sections in the book before you move on. You could even use this problem set as a practice final exam for your own interest. These problems will give you feedback on whether you remember the math that you learned in each chapter of this book.

Look Back–Comprehensive Review for Master Math's Mysteries, Chapters 1-10

Perform the indicated operation:

1. $8 \div (-4)$ **2.** $-11 \cdot (-5)$

3. $3 - (-2)$ **4.** $-2 + (-5)$

Fill in the blank:

5. The product of 9 and 4 is _____.

6. The sum of 32 and 6 is _____.

7. The sum of 9 and -4 is _____.

8. The difference of -2 and 6 is _____.

Perform the indicated operation:

9. $-3 - (-4)$ **10.** $(-25) \div (-5)$

11. $7(25)$ **12.** $5 + (-9)$

Try these "order of operation" problems:

13. $9(2) - 4$ **14.** $13 + 3(8 - 4)$

Perform the indicated operation:

15. $\dfrac{5}{6} - \dfrac{1}{6}$ **16.** $220 + 230$

17. $8.1 - 4.2$ **18.** $0.005 + 21.132$

19. $\dfrac{1}{3} \cdot \dfrac{2}{5}$ **20.** $\dfrac{1}{6} + \dfrac{1}{3} + \dfrac{1}{4}$

CHAPTER 12

Optional: Consider Algebra

"Life shrinks or expands according to one's courage."

ANAÏS NIN

Introduction

Algebra is a gatekeeper to many exciting majors and professions. Successfully completing algebra opens doors for you. The goal of this chapter is not to teach you algebra but to give you a taste of the basics of algebra and a framework for your further learning. Hopefully it will be a chapter that will encourage you to continue with your education in mathematics. Dr. David Drew, research/statistics professor, says, "We must recognize that virtually every student, regardless of ethnicity or gender can master mathematics. . . . We should expect and require them to do so" (Drew, 1996, p. 12). Please expect and require yourself to do everything possible to learn the math you need to keep your options open in the careers you pursue during the 40 years you will be in the workforce.

Algebra or Arithmetic?

Algebra and arithmetic are different in one important way. They both allow you to add, subtract, multiply, and divide, but algebra allows you to add, subtract, multiply, and divide *with numbers that you don't know*. The favorite name for those numbers we don't know is "*x*." The second favorite name is "*y*," but any letter or symbol could be used as a temporary replacement for the number we don't know, which is also referred to as the "unknown." The letter "n" is also a favorite substitute for the unknown number. These letter substitutes are called "variables" because what they represent changes or varies. In some situations, we are able to figure out the unknown number (that is, figure out the x) and, in other situations, we are not.

When We Say	We Mean
$3x$	3 times some number we don't know
$4 - x$	4 take away some number we don't know
$7 + 2x$	7 plus 2 times some number we don't know
x^2 or x squared	Some number we don't know times itself
$x = 2 + 5$	The number we don't know is 2 plus 5

See if you can translate the symbols in the left-hand column into the words they mean. (Answers are in the Appendix.)

When We Say	We Mean
$x + 10$	1.
$6x - 2$	2.
$\dfrac{5}{x}$	3.
$4 - 2 = x$	4.

Expression or Equation?

To do well in algebra, you must learn certain basic concepts and vocabulary words. Two of the most essential words to understand are "expression" and "equation." Many people incorrectly use "equation" loosely to refer to anything that looks like algebra. They confuse equations with expressions. But an expression and an equation are very different, and it is important to know the difference. There are algebra procedures that can be performed only on an expression and other procedures that can be performed only on an equation.

■ An **expression** is numbers and/or variables that are usually operated on by addition, subtraction, multiplication, or division. Here are five examples of expressions. Notice that none of these expressions has an "$=$" sign.

$$3x \qquad 4 + n \qquad 6y - 3 \qquad \frac{5 - x}{3} \qquad 8 - 5$$

■ An **equation** is formed when two expressions are set equal. An equation is a left-side expression set equal to a right-side expression. All equations have two sides—a left side and a right side—separated by an equals sign. $3 + 2 = 5$ is an equation. $3 + 2$ is the left side and 5 is the right side. Here are five examples of equations:

$$n = 4 - 7 \qquad x + 3 = 5 \qquad 4x = 16 \qquad 3 = 2(x - 1) \qquad 5 = 2 - y$$

Notice that each of these equations has one "$=$" sign. Each of them is a left-side expression equal to a right-side expression.

Left Side	Equals	Right Side
n	=	$4-7$
$x+3$	=	5
$4x$	=	16
3	=	$2(x-1)$
5	=	$2-y$

Identify each of the following as either an expression or an equation. (See the answers in the Appendix.)

5. $y+3$ **6.** $4x$ **7.** $n=2$ **8.** $2-x$

9. 5 **10.** $5=3n$ **11.** $2+4=x$ **12.** $\dfrac{5}{y}$

13. $x=\dfrac{5}{2}$ **14.** $3x-7$ **15.** $6+4n$ **16.** $5=y$

The "Big Five" Procedures in Algebra

For most of beginning algebra, five procedures are used again and again. These five important procedures are simplify, factor, evaluate, solve, and graph. The solutions of almost all algebra problems involve these procedures. **We simplify, factor, and evaluate expressions. We solve and graph equations.** These five procedures, what they mean, and examples of how to perform each one summarize the work beginning algebra students do in one semester. The rest of this chapter will introduce you to these five procedures and give you examples of what they mean.

This song, sung to the tune of the Christmas carol "O Come, All Ye Faithful," can help you remember the five important algebra procedures.

I Love My Algebra ♪

I love my algebra.

Morning, noon, and night,

I do algebra whenever there is light.

Simplify and solve all that I can take.

Graph, factor, an' evaluate.

Graph, factor, an' evaluate.

Graph, factor, an' evaluate,

They're a piece of cake.

(Lyrics by Mike Petyo and Cheryl Ooten)

Procedure #1: Simplify

When you simplify your life, you cut back. You may cut back on your activities or clutter in your home. In algebra, "simplify" means the same—cut back on activities or "math clutter." Here are some examples of simplifying expressions in algebra:

Algebraic Expression	Simplification
$25 + 75$	100
$2 + x + 5$	$7 + x$
$x - 3 - 4$	$x - 7$
$3(x + 4)$	$3x + 12$
	Note: We are multiplying 3 times the quantity of some number we don't know plus 4. Because we can't add a number we don't know plus 4, the order of operations (Chapter 3's "Master Math's Mysteries") says to multiply by 3. That can be done by multiplying the 3 times some number and adding 3 times 4.
$2x - 5 + 7$	$2 + 2x$

Simplify the expressions in the following table. (See the answers in the Appendix.)

Expression	Simplified Expression	Expression	Simplified Expression
17. $3 + 5 - 6$		18. $2 + 3(9)$	
19. $x + 5 + 3$		20. $3 + 7 + 2x$	
21. $x + 4 + 5$		22. $4 \div 2 + 6(8 - 1)$	

One of the concepts that we learn as we simplify expressions is "like terms." Like terms involve adding or subtracting expressions that have the same "number we don't know." For example, $2x$ and $5x$ involve the same unknown number, but $2x$ and $5y$ do *not* involve the same unknown number. So, $2x$ and $5x$ are like terms, whereas $2x$ and $5y$ are not. Because $2x$ and $5x$ are like terms, we can add or subtract them. For example: $2x + 5x$ means "2 times some number we don't know plus 5 times that same number." Instead of saying "$2x + 5x$," it would be simpler to say "$7x$."

Here are some examples of simplifying expressions in algebra by "combining like terms" followed by some problems for you to practice combining like terms on your own.

    a. $6x + 7x$ simplifies to $13x$.

    b. $6y + 7x$ cannot be simplified because they are not like terms.

    c. $8x - 5x$ simplifies to $3x$.

    d. $2y - 10y$ simplifies to $-8y$. (Recall signed numbers in Chapter 7's "Master Math's Mysteries.")

Practice combining like terms in this table. (See the answers in the Appendix.)

Expressions	Simplified Expressions	Expressions	Simplified Expressions
23. $2x + 7x$		24. $9x - 4x$	
25. $3y + 4y$		26. $8y - 2x$	
27. $2n + 5n$		28. $6x - 9x$	

Procedure #2: Factor

The procedure "factor" means to turn an expression into a multiplication problem. Then those parts of the multiplication problem that are multiplied are called "factors." In basic math, you learned to factor numbers such as 10. By changing 10 into 2 times 5, you "factored 10." The numbers 2 and 5 are called factors. Recall that we often show multiplication by using parentheses. We can factor 10 and write it as 2 times 5 or 2(5) or (2)(5) or 2 · 5. We rarely write "2 × 5" in algebra because the × symbol for multiplication looks too much like our letter x, which we substitute for the number that we don't know. To avoid confusion, we use either parentheses or the raised dot for multiplication.

   You might also have learned to factor larger numbers such as 24 by making a factor tree. Here are three factor trees that show that 24 factors as 2(2)(2)(3) or $2^3 \cdot 3$.

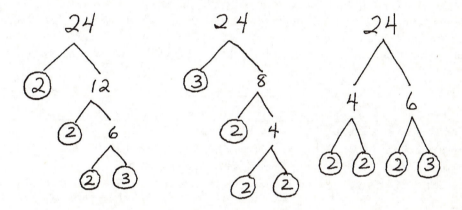

Notice that we placed 24 at the top of the tree and then factor 24 any way we wished—2(12) or 3(8) or 4(6)—making branches to each of those factors. Then we factored each new factor again when possible. When we reached the end of a branch and could factor no further, we circled the number to remember to keep it. Notice that in all three factor trees, 2, 2, 2, and 3 were at the ends of the final branches. That means that 24 factors into the unique product 2(2)(2)(3). The numbers 2 and 3 are called **prime numbers** because they cannot be factored again in an interesting way. (Note that 2 is 2 · 1 or 1 · 2 and 3 is 3 · 1 or 1 · 3 but we don't consider those interesting.) The number 24 is called a **composite number** because it factors in an interesting way. Twenty-four factors as 2(2)(2)(3) or 2 · 2 · 2 · 3 or $2^3 \cdot 3$, which is called the **product of prime factors**.

   Practice the factoring procedure by factoring the following numbers using a factor tree. Place the number to be factored at the top of the tree. Then make two branches with any two factors at the lower ends that multiply to the top number. Keep making branches with the new numbers until you reach prime numbers (that is, they cannot factor again in an interesting way) and circle the primes. Collect the prime numbers at the ends of the branches and write them as a product. That product is your answer.

29. Factor 15   **30.** Factor 50

   In algebra, we factor expressions. We do not factor equations. Here are some examples of factoring expressions in algebra:

   $4x^2$ can be written as $(4)(x^2)$ or as $(2)(2)(x)(x)$ with prime factors.

   $12y^3$ can be written as $(12)(y^3)$ or as $2 \cdot 2 \cdot 3 \cdot y \cdot y \cdot y$ with prime factors.

When you have had more experience with algebra than we can provide here, you will learn to factor more complicated algebraic expressions. One example of factoring a more involved expression is that $4x - 6$ factors as $2(2x - 3)$. Do not worry why that is true right now. Just notice that it factored into a multiplication problem: 2 times $(2x - 3)$, which is one expression times another. A second more difficult example is that $x^2 + 2x$ factors as $(x)(x + 2)$. This answer is the expression x times the expression $(x + 2)$. I repeat my request that you do not worry why these two examples are true now. These two examples require more explanation and experience with algebra. You might recognize that in the two examples we changed a difference into a product and then a sum into a product.

Procedure #3: Evaluate

To "evaluate" means "to find the value of an expression." Notice that "*evaluate*" contains the letters "valu," which remind you that "evaluate" means to find the "value." To evaluate "2 + 3" means to add to a value of 5. If you are asked to evaluate an expression containing an unknown such as $3x$, you must be told the value of the unknown before you can evaluate it.

Examples	Solutions
(a) Evaluate $x + 6$ if x stands for 7.	$7 + 6$ is 13. The value is 13.
(b) Evaluate $3x$ if x stands for 4.	$3x$ is 3 times x. So, if x is 4, $3x$ is 3(4) or 12. The value is 12.
(c) Evaluate $5x$ if x is -2.	5 times -2 is -10. The value is -10.
(d) Evaluate $6 - x$ if x is -3.	$6 - (-3)$ is "6 light beans take away 3 dark beans" or 9. The value is 9. (See Chapter 7's "Master Math's Mysteries.")
(e) Evaluate $\dfrac{12}{x}$ if $x = -4$.	$\dfrac{12}{-4} = -3$ The value is -3.
(f) Evaluate $2x - 3$ if $x = 6$.	Order of operations (PEMA—See Chapter 3's "Master Math's Mysteries") tells us to multiply before subtracting. So $2x$ becomes 2 times 6 or 12. Then $12 - 3$ is 9 and the value is 9.

Here are some expressions for you to evaluate. (See the answers in the Appendix.)

Problem	Solution
31. Evaluate $5x$ when x stands for 6	
32. Evaluate $7 + x$ when x is -3	
33. Evaluate $x - 5$ when x is -2	
34. Evaluate $2x + 8$ when $x = 3$	
35. Evaluate $8 - x$ when $x = -2$	

Did you notice?

We have used Procedures 1, 2, and 3 (Simplify, Factor, and Evaluate) *only* on expressions—not on equations.

Procedures 4 and 5 (Solve and Graph) will be used on equations—not expressions.

Procedure #4: Solve

We often think of detectives solving mysteries or answering questions. In algebra, solving those mysteries involves finding an unknown number. The unknown number might be a person's age or the cost of a car or the number of miles traveled. We can find the unknown number **only if** we are given enough information. Often that information is either given in the form of an equation or in the form of words that can be translated into the form of an equation. (Remember that an equation is a left-hand expression equal to a right-hand expression.) In fact, if it is possible to discover the unknown number, we are often told to "solve the equation" or "solve the word problem."

Solving Equations.   Here are examples of equations being solved:

Example 1

Solve the equation: $\qquad\qquad\qquad$ $x = 8 - 13$
Simplify right side and keep left side the same: $\quad x = -5$
Now we know the unknown number is -5.

Example 2

Solve the equation: $\qquad\qquad\qquad\qquad$ $2x = 24$
Think: 2 times some number gives me 24. $\qquad (2)(?) = 24$
The missing number must be 12 because
$\quad$ 2 times 12 is 24.
So $\qquad\qquad\qquad\qquad\qquad\qquad\qquad$ $x = 12$
and we know the unknown number is 12.

Example 3

Solve the equation: $\qquad\qquad\qquad\qquad$ $3 + x = 18$
Think: 3 plus some number gives me 18. $\qquad 3 + ? = 18$
The missing number must be 15 because
$\quad$ 3 plus 15 is 18. $\qquad\qquad\qquad\qquad\qquad 3 + 15 = 18.$
So $\qquad\qquad\qquad\qquad\qquad\qquad\qquad$ $x = 15$
and we know the unknown number is 15.

See if you can solve the following equations by just using your common sense to discover what unknown number would make your equation true. (Answers are in the Appendix.)

36. Solve the equation: $5x = -40$

37. Solve the equation: $x + 7 = 19$

38. Solve the equation: $x - 4 = 22$

You may have been able to solve these equations using common sense or guessing. Unfortunately, most equations are not solved as simply. When you take a formal algebra course, you will learn logical steps that will help you solve the more complicated equations. An important guideline for solving equations is that an equation must stay **balanced**. That means, "Whatever you do to one side of an equation (add, subtract, multiply, or divide), you must do the identical operation (add, subtract, multiply, or divide) on the other side."

For example:

When Solving an Equation, If You	Then You Must
Add 3 to the left side	Add 3 to the right side.
Subtract 5 from the right side	Subtract 5 from the left side.
Multiply the left side by 4	Multiply the right side by 4.
Divide the right side by 2	Divide the left side by 2.

When you take a prealgebra or beginning algebra course, you will get lots of practice balancing equations as you solve them.

Solving a Word Problem Using Algebra.   Chapter 9 gave you "tricks" or strategies for solving word problems, which included the strategies "use algebra" and "guess and check." We will work through the following word problem first using the guess and check problem-solving strategy and then using algebra that comes from the understanding of the problem we gained by guessing and checking first. We will use Polya's problem-solving steps from Chapter 9.

THE DRIVING PROBLEM

Don will be driving west at 75 miles per hour while Luisa drives east at 60 miles per hour. If they begin at the same place, how long will it take them to be 405 miles apart?

We need to find the amount of time it takes for them to be 405 miles apart. Let's draw what's happening and use the guess and check strategy to understand the problem more (see the drawing on the next page). They are going in opposite directions, so adding the distances that they each travel will give us the number of miles they are apart. After just one hour, Don has driven 75 miles and Luisa has driven 60 miles. (That's what 75 mph and 60 mph mean.) Each hour, Don drives 75 miles farther and Luisa drives 60 miles farther. The three diagrams show what happens each hour. After one hour, they are 135 miles apart. After two hours, the total of their two distances traveled is 270 miles. After three hours, the total of Don's 225 miles and Luisa's 180 miles is 405 miles. They are 405 miles apart—our goal. We have guessed three times—one hour, two hours, and three hours—and find the answer to be three hours.

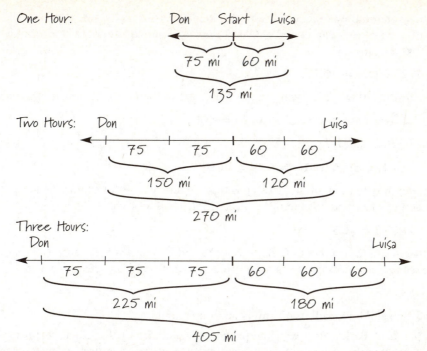

Reworking the Driving Problem Using Algebra.   Reread the Driving Problem. Remember that we know the answer is three hours. Let's pretend we don't know this. This time let's use a letter (a variable) to represent the answer to our problem. We will practice Polya's four steps from Chapter 9 here too.

Step 1. Understand the Problem.   We know that Don and Luisa drive opposite directions. Their speeds tell us how far they will travel each hour. Don will travel 75 miles every hour. Luisa will travel 60 miles every hour. We want to find the number of hours it will take them to be 405 miles apart. So, we write down:

> Find the time it takes Don and Luisa to be 405 miles apart.

Step 2. Devise a Plan.   We plan to use algebra. Let's represent the number of hours with the letter t for time. We write down:

> Let t = time (in hours)

Using the diagrams that we drew previously, we see that our goal is:

> Don's distance + Luisa's distance = 405 miles

This is called an equation (left side = right side). Also notice from our previous diagram that Don's distance is the number of hours times 75 and that Luisa's distance is the number of hours times 60. Because t represents the number of hours,

> Don's distance = $t \cdot 75$ and Luisa's distance = $t \cdot 60$

so the equation becomes:

> $t \cdot 75 + t \cdot 60 = 405$

Step 3. Carry Out the Plan.   We have a plan, and it is to solve the equation to figure out what the letter t would be. Don't worry about solving; I will walk you through it. In solving an equation, the left side always has to balance the right side.

> $t \cdot 75 + t \cdot 60 = 405$

When you multiply, order doesn't matter. Remember that 2 • 3 = 3 • 2 = 6, so $t • 75 = 75 • t$ or 75t and $t • 60 = 60 • t$ or 60t, and our equation can be rewritten this way:

$$75t + 60t = 405$$

When you multiply 2 • 3, you can think of it as two threes, so our equation is really:

seventy-five *t*'s plus sixty *t*'s = 405

which is the same as:

one hundred thirty-five *t*'s = 405

Another way to think of the last step is that 75t and 60t are "like terms," and we have combined like terms 75t and 60t to make 135t. Then we have:

$$135 • t = 405$$

We need a number that multiplies times 135 and becomes 405. Because 135 • 3 is 405, we see that the unknown *t* has to be equal to 3. In algebra language, we write:

$$t = 3$$

And Don and Luisa will be 405 miles apart in 3 hours.

Step 4. Look Back.   Reviewing our previous diagram, we see that our answer makes sense and checks in the original problem. Consider these questions. Is algebra a method you might use again to solve another problem using distances? Do you know that distance is equal to rate times time? Is there any place in this problem where that applies?

Procedure #5: Graph

A graph is a picture of the data, which are often numbers. For most people, a picture increases understanding and makes the relationships between the numbers more clear than just viewing a list of numbers. There are many kinds of graphs. Two common types of graphs—bar graph and line graph—illustrate the data in the table. The data in the table show quarterly earnings for the Mesa Verde Bookstore. Notice how the graphs give an interesting picture showing trends not readily noticeable by just looking at the numbers in the chart.

	First Quarter	Second Quarter	Third Quarter	Fourth Quarter
Earnings	$5,800	$10,300	$8,500	$4,700

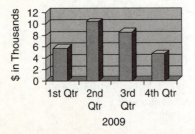

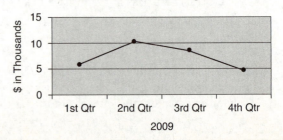

The Cartesian Coordinate System

In algebra, we often graph on something called the Cartesian coordinate system. Recall Story #1 from Chapter 9 about René Descartes, who saw a fly on the ceiling and realized that the fly's position could be described by two numbers that are the perpendicular distances from the edges of the ceiling. Descartes imagined his ceiling to look like Figure 12.1.

Let's expand on Descartes's vision of the ceiling. Imagine that we lie on our backs on the floor and can now see flies on the ceilings in three other rooms that join ours at one common corner. Let's also pretend that the walls between rooms suddenly transform into single straight lines along the ceiling and that they become infinite in length (suggested by arrows). See Figure 12.2.

Let's place some flies on those ceilings and describe their locations. Recall Chapter 10's "Master Math's Mysteries," where the subway train station became the zero point on our number lines. Because we now need a zero point to begin measuring for two number lines, the point that works for all four ceilings is the common corner. We call that common corner the "origin," and the location of any fly that lands there is (0, 0). (See Figure 12.2.) That means the fly is zero distance from the origin in any direction: left/right or up/down.

Figure 12.1

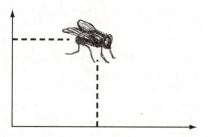

Figure 12.2

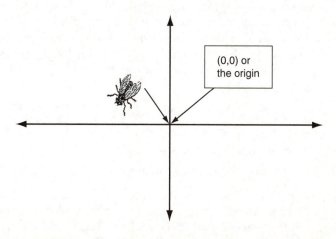

(0,0) or the origin

Figure 12.3

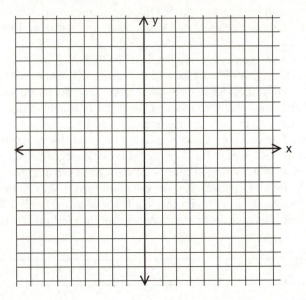

Remembering that we are lying on the floor looking up at the ceiling, we call the horizontal edge the x-axis and the vertical edge the y-axis. Today our Cartesian coordinate system (named after Descartes) looks like Figure 12.3.

Because we need to know whether to move left or right and whether to move up or down, let's agree that the signs of the distances we use will again determine direction. Let's agree that when we're moving horizontally (left or right), a negative number means to move left from the origin and a positive number means to move right (like our Chapter 10 number lines). Let us also agree that when we're moving vertically (up or down), a negative number means to move down and a positive number means to move up. Because the left/right and up/down numbers will usually be different, let us also agree to pair the numbers in parentheses and to give the left/right (horizontal) number first before the up/down (vertical) number. Writing a number pair such as $(-2, 5)$ would mean that our fly is 2 units left of the origin and 5 units up from the origin (the common corner of the ceilings). Locating that fly, we could begin at the origin and walk left (if we were flies) 2 units and then turn right, going up 5 units. See Figure 12.4. (This is almost like walking in a neighborhood where all the blocks are square and of the same size.)

Practice by finding the number pairs that locate the four flies (indicated by dots) in Figure 12.5. Then check your number pairs with my answers.

The fly locations in Figure 12.5 are:

1. $(3, 4)$ From the origin, go right 3 and up 4.
2. $(-1, 3)$ From the origin, go left 1 and up 3.
3. $(-4, -2)$ From the origin, go left 4 and down 2.
4. $(1, -2)$ From the origin, go right 1 and down 2.

Although his discovery paved the way for the Cartesian coordinate system, which we use in algebra today, René Descartes would not use negative numbers because he was studying distances and was not open to consider the negative sign as a change of direction. He even said that negative solutions of equations were "false" solutions (Berlinghoff & Gouvea, 2004, p. 98).

Figure 12.4

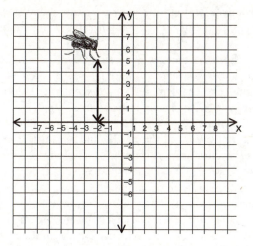

Figure 12.5

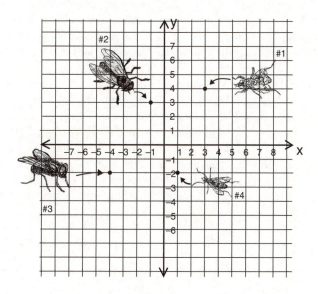

Now, of course, the Cartesian coordinate system would not still be used if its only use were to locate bugs on the ceiling. As we said at the beginning of this section, a graph is a picture of the data and in algebra we have lots of data to place on the Cartesian coordinate system.

Graphing Data from Equations

Let's consider the equation $y = 2x$. Notice that it is an equation because it has a left side, y, equal to a right side, $2x$. Notice also that it has two unknowns, y and x. When we say $y = 2x$, we mean we want to look at all numbers (symbolized by y) that are twice some other number (symbolized by x). Think about the possibilities. The number 4 is twice 2. The number 10 is twice 5. The number -2 is twice -1. We could continue on and on to finding pairs of numbers where one of them (we called it y) is twice the other (we called it x) but that would take too much time. The advantage of graphing is that we can actually see a picture of all those number pairs even though we could never finish calculating them all because there are infinitely many.

So we will organize the pairs that we have already found. One of the problem-solving strategies from Chapter 9 was to organize the data. You may recall that we organized our number pairs in the Handshake Problem by using a t-table. Let's do that now and add a few more number pairs where y is twice x. See Figure 12.6.

Notice in the t-table in Figure 12.6 that I have put the x-part of the pairs on the left and the y-part of the pairs on the right. We call the x-value of the pair the **x-coordinate**. We call the y-value the **y-coordinate**. Because our equation says, "y is twice x," we notice that the y-value actually depends on the x-value. What we know about y is that y is twice another number we called x. We really don't know anything about the x, so we say it is independent and we call it the "independent variable." But because we do know that "y is twice x" and because y depends on x, we call y the "dependent variable." In our organizing t-tables, we place the independent variable on the left and the dependent variable on the right.

We extended our t-table for $y = 2x$ and could continue extending it indefinitely. This is where a graph (or picture of the data) is very useful. For reasons beyond us at this level of algebra and for lack of space in this book, I am going to ask you to believe that the equation $y = 2x$ is a "linear equation," which means that the infinite number pairs in which the second one of them is twice the first lie on a straight line when positioned on Descartes's Cartesian coordinate system. If we use the number pairs in our t-table as locations of flies, all of those flies would be standing in a straight line. (Wouldn't that be fun to see?) See Figure 12.7, in which all the number pairs from the t-table are positioned, and notice that those dots make a straight line.

Because I have told you that $y = 2x$ is a linear equation, we can draw the straight line through all of the fly locations and extend it infinitely in both directions. See Figure 12.8. The amazing thing about this picture (called a graph) is that it is a picture of all the number pairs where y is twice x. Also, every fly that lands on this line will have its location described by a pair of numbers in which one of them is twice the other. Being able to look at all of the number pairs that fit an equation was the great gift that Descartes gave the world when he lay on his back watching the fly walk the ceiling. His discovery brought about a visual way—the graph on the Cartesian coordinate system—to view the infinite number of solutions to equations, such as $y = 2x$, in algebra.

Figure 12.6

x	y
2	4
5	10
−1	−2
1	2
3	6
−2	−4

Figure 12.7

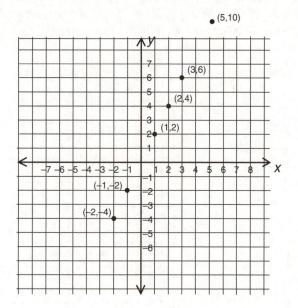

Figure 12.8

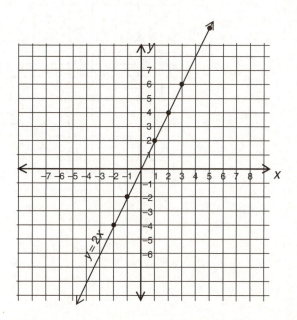

You may wish to pick one of the following equations and find number pairs (like fly locations) that would fit the equation. If you find at least three number pairs, you could make a straight-line picture of all the solutions (number pairs that work) of that equation. I have included t-tables so you can organize your number pairs and Cartesian coordinate systems if you would like to try them. Solutions are in the Appendix.

39. $y = x + 2$ Hint: Look for number pairs in which one number (y) is two more than the other (x).

x	y
3	
1	
0	
−5	

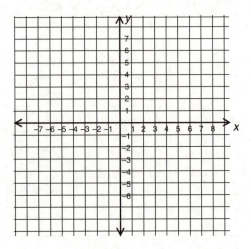

40. $y = 2x + 1$ Hint: Look for number pairs in which one number (y) is one more than twice the other (x).

x	y
0	
1	
3	
−2	

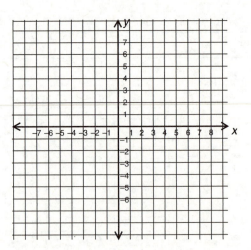

41. $y = x - 1$ Hint: This time pick any values for x and place them in the left column. Look for number pairs in which one number (y) is one less than the other (x).

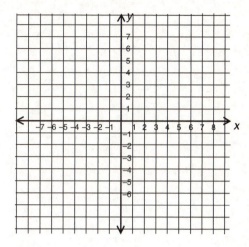

Conclusion

You have now experienced the five main procedures taught in algebra—simplify, factor, evaluate, solve, and graph. You also learned important vocabulary—expression and equation. When you study algebra, keep this big picture in mind and you will go far. Remember that you simplify, factor, and evaluate expressions, whereas you solve and graph equations.

If you have not followed everything in this chapter, it is not because of you. I have tried to summarize what I consider the most important ideas in algebra. Summaries are often more difficult to understand than what is being summarized. As you continue your study of algebra, pay close attention to the difference between expressions and equations and to the five procedures (simplify, factor, evaluate, solve, and graph). Remember what researcher/statistics professor David Drew (1996) says: "Virtually everyone can learn advanced mathematical concepts, even those who start late."

A P P E N D I X

Bibliography

Introduction

Bailey, Elinor Peace. "Let's Face It." Piecemakers, Costa Mesa, CA. 18 April 1999.

———. *Mother Plays with Dolls . . . and Finds an Important Key to Unlocking Creativity.* McLean, VA: EPM Publications, 1990.

Bandler, Richard. *Using Your Brain for a Change.* Moab, UT: Real People Press, 1985.

Bandler, Richard, and John Grinder. *Frogs into Princes.* Moab, UT: Real People Press, 1979.

Beck, Aaron, M.S., and Gary Emery, Ph.D. *Anxiety Disorders and Phobias: A Cognitive Perspective.* New York: Basic Books, 1985.

Cooney, Miriam P., csc, ed. *Celebrating Women in Mathematics and Science.* Reston, VA: National Council of Teachers of Mathematics, 1996.

Goldberg, Natalie. *Writing Down the Bones.* Boston: Shambhala, 1986.

Greenberger, Dennis, Ph.D., and Christine A. Padesky, Ph.D. *Mind over Mood: Change How You Feel by Changing the Way You Think.* New York: Guilford Press, 1995.

Hackworth, Robert D. *Math Anxiety Reduction,* 2nd ed. Clearwater, FL: H&H Publishing, 1992.

Kass-Simon, G., and Patricia Farnes, ed. *Women of Science: Righting the Record.* Bloomington, IN: Indiana University Press, 1990.

LeDoux, Joseph. *The Emotional Brain.* New York: Touchstone, 1996.

McGrayne, Sharon Bertsch. *Nobel Prize Women in Science: Their Lives, Struggles and Momentous Discoveries.* New York: Birch Lane Press, 1993.

Meichenbaum, Donald. DVD. *Mixed Anxiety and Depression: A Cognitive-Behavioral Approach.* San Francisco: Psychotherapy.net, 2006.

Ooten, Cheryl. Address, Santa Ana College commencement. Santa Ana Stadium, Santa Ana, CA. 4 June 1999.

Tobias, Sheila. *Overcoming Math Anxiety.* New York: W. W. Norton, 1993.

Chapter 1

Adair, Margo. *Working Inside Out: Tools for Change.* Berkeley, CA: Wingbow Press, 1984.

Andreas, Brian. *Mostly True.* Decorah, IA: StoryPeople, 1993.

Bandler, Richard. *Using Your Brain for a Change.* Moab, UT: Real People Press, 1985.

Bandler, Richard, and John Grinder. *Frogs into Princes.* Moab, UT: Real People Press, 1979.

Berlinghoff, William P., and Fernando Q. Gouvea. *Math Through the Ages: A Gentle History for Teachers and Others*. Farmington, ME: Oxton House and Mathematical Association of America, 2004.

Clawson, Calvin C. *Mathematical Mysteries: The Beauty and Magic of Numbers*. New York: Plenum Press, 1996.

Day, Caren. *Mental Math: Mini Mental Workouts to Improve Computation Skills*. Huntington Beach, CA: Creative Teaching Press, 2001.

Dilson, J. *The Abacus: The World's First Computing System*. New York: St. Martin's Griffin, 1968.

Gardner, Howard. *Multiple Intelligences: The Theory in Practice*. New York: Basic Books, 1993.

Green, P. *How to Use a Chinese Abacus*. Morrisville, NC: Lulu, 2007.

Greenberger, Dennis, Ph.D., and Christine A. Padesky, Ph.D. *Mind over Mood: Change How You Feel by Changing the Way You Think*. New York: Guilford Press, 1995.

Gross, Ronald. *Peak Learning*. New York: Tarcher/Putnam, 1991.

Hart, Leslie. *Human Brain & Human Learning*. Kent, WA: Books for Educators, 1998.

Hemenway, P. *Divine Proportions: Phi in Art, Nature, and Science*. New York: Sterling Publishing, 2005.

Hersh, Reuben. *What Is Mathematics, Really?* New York: Oxford University Press, 1997.

Hope, Jack A., Barbara Reys, and Robert Reys. *Mental Math in Junior High*. Palo Alto, CA: Dale Seymour Publications, 1988.

Kogelman, Stanley, and Barbara R. Heller. *The Only Math Book You'll Ever Need*. New York: HarperCollins, 1994.

Kogelman, Stanley, and Joseph Warren. *Mind over Math*. New York: McGraw-Hill, 1978.

LeDoux, Joseph. *The Emotional Brain*. New York: Touchstone, 1996.

National Council of Teachers of Mathematics. *Historical Topics for the Mathematics Classroom*. Washington, DC: National Council of Teachers of Mathematics, 1969.

Tang, Greg. *The Grapes of Math: Mind-Stretching Math Riddles*. New York: Scholastic Press, 2001.

Tobias, Sheila. *Overcoming Math Anxiety*. New York: W. W. Norton, 1993.

Chapter 2

Csikszentmihalyi, Mihaly. *Creativity: Flow and the Psychology of Discovery and Invention*. New York: Harper Perennial, 1996.

————. Creativity & Innovation Course. Claremont, CA: Claremont Graduate University, Spring Semester, 2007.

————. *Finding Flow*. New York: Basic Books, 1997.

————. Managing Flow in Organizations Course. Claremont, CA: Claremont Graduate University, Spring Semester, 2007.

DePorter, Bobbi, and Mike Hernacki. *Quantum Learning: Unleashing the Genius in You*. New York: Dell, 1992.

Hart, Leslie. *Human Brain & Human Learning*. Kent, WA: Books for Educators, 1998.

Chapter 3

Albers, Donald J., Gerald L. Alexanderson, and Constance Reid, eds. *More Mathematical People: Contemporary Conversations*. San Diego, CA: Academic Press, 1990.

Bandler, Richard. *Using Your Brain for a Change*. Moab, UT: Real People Press, 1985.

Clawson, Calvin C. *Mathematical Mysteries: The Beauty and Magic of Numbers*. New York: Plenum Press, 1996.

DePorter, Bobbi, and Mike Hernacki. *Quantum Learning: Unleashing the Genius in You*. New York: Dell, 1992.

Drew, David. *Aptitude Revisited: Rethinking Math and Science Education for America's Next Century*. Baltimore, MD: Johns Hopkins University Press, 1996.

Gardner, Howard. *Multiple Intelligences: The Theory in Practice*. New York: Basic Books, 1993.

Gullberg, Jan. *Mathematics: From the Birth of Numbers*. New York: W. W. Norton, 1997.

Kass-Simon, G., and Patricia Farnes, eds. *Women of Science: Righting the Record*. Bloomington, IN: Indiana University Press, 1990.

McGrayne, Sharon Bertsch. *Nobel Prize Women in Science: Their Lives, Struggles, and Momentous Discoveries*. New York: Birch Lane Press, 1993.

Morrison, Philip, Phylis Morrison, and the Office of Charles and Ray Eames. *Powers of Ten*. New York: Scientific American Library, 1994.

Nelson, David, George Gheverghese Joseph, and Julian Williams. *Multicultural Mathematics*. New York: Oxford University Press, 1993.

Pappas, Theoni. *The Magic of Mathematics*. San Carlos, CA: Wide World Publishing/Tetra, 1996.

Reid, Constance. *Julia: A Life in Mathematics*. Washington, DC: Mathematical Association of America, 1996.

Shorris, Earl. *Latinos: A Biography of the People*. New York: Avon Books, 1992.

Stewart, Ian. *Another Fine Math You've Got Me Into . . .* New York: W. H. Freeman, 1992.

Sunbeck, Deborah. *Infinity Walk: Preparing Your Mind to Learn!* Torrance, CA: Jalmar, 1996.

Chapter 4

Csikszentmihalyi, Mihaly. Creativity & Innovation Course. Claremont, CA: Claremont Graduate University, Spring Semester, 2007.

Student/Tutor Interviews, Santa Ana College, June 2001.

Chapter 5

Student/Tutor Interviews, Santa Ana College, June 2001.

Chapter 6

Hart, Leslie A. *Human Brain & Human Learning*. Kent, WA: Books for Educators, 1998.

McMeekin, Gail. *The 12 Secrets of Highly Creative Women*. New York: MJF Books, 2000.

Schrof, Joannie M., and Stacey Schultz. "Social Anxiety." *U.S. News & World Report,* 21 June 1999.

Sousa, David, Ed.D. "Seven Critical Discoveries About How the Brain Learns," Brain Expo 2000. Paradise Point Hotel, San Diego, CA. 18 January 2000.

Wolfe, Pat. "Lessons Learned from Brain Research," Brain Expo 2000. Paradise Point Hotel, San Diego, CA. 18 January 2000.

Chapter 7

Benson, Herbert, M.D. *The Relaxation Response*. New York: Avon, 1975.

Berlinghoff, William P., and Fernando Q. Gouvea. *Math Through the Ages: A Gentle History for Teachers and Others*. Farmington, ME: Oxton House and Mathematical Association of America, 2004.

Burns, David D., M.D. *Feeling Good: The New Mood Therapy*. New York: Avon, 1999.

Engelbreit, Mary. Card. Bloomington, IN: Sunrise, 1996.

Nelson, David, George Gheverghese Joseph, and Julian Williams. *Multicultural Mathematics: Teaching Mathematics from a Global Perspective*. Oxford, UK: Oxford University Press, 1993.

Pappas, T. *The Magic of Mathematics: Discovering the Spell of Mathematics*. San Carlos, CA: Wide World Publishing, 1996.

Chapter 8

Bromley, Karen, Linda Irwin-DeVitis, and Marcia Modlo. *Graphic Organizers: Visual Strategies for Active Learning.* New York: Scholastic, 1995.

Buzan, Tony, with Barry Buzan. *The Mind Map Book: How to Use Radiant Thinking to Maximize Your Brain's Untapped Potential.* New York: Dutton, 1994.

Cahill, Larry, Ph.D. "Emotions and Memory." Brain Expo 2000, Paradise Point Hotel, San Diego, CA. 19 January 2000.

Grafton, Sue. *O Is for Outlaw.* Markham, Ontario, Canada: Henry Holt, 1999.

Hestwood, Diana. "Unraveling the Mysteries of the Emerging Adult Brain Ages 18–20." American Mathematical Association for Two Year Colleges Annual Conference. Cincinnati, OH. 3 November 2006.

Margulies, Nancy, and Nusa Maal. *Mapping Inner Space: Learning and Teaching Visual Mapping,* 2nd ed. Chicago: Zephyr Press, 2002.

Markowitz, Karen, M.A., and Eric Jensen, M.A. *The Great Memory Book.* San Diego, CA: The Brain Store, 1999.

Rossi, Ernest Lawrence, Ph.D., with David Nimmons. *The 20 Minute Break.* Los Angeles: Jeremy P. Tarcher, 1991.

Sapolsky, Robert, Ph.D. "Stress, Disease, and Memory." Brain Expo 2000. Paradise Point Hotel, San Diego, CA. 18 January 2000.

Siegel, Daniel, M.D. "The Developing Mind." Brain Expo 2000. Paradise Point Hotel, San Diego, CA. 18 January 2000.

Wycoff, Joyce. *Mindmapping: Your Personal Guide to Exploring Creativity and Problem-Solving.* New York: Berkley, 1991.

Zola, Dr. Stuart. "Memory Seminar." Institute for Cortext Research and Development Seminars. Marriott Hotel, Anaheim, CA. 20 September 1999.

Chapter 9

Afflack, Ruth. *Beyond Equals: To Encourage the Participation of Women in Mathematics.* Oakland, CA: Math/Science Network, 1982.

Berlinghoff, William P., and Fernando Q. Gouvea. *Math Through the Ages: A Gentle History for Teachers and Others.* Farmington, ME: Oxton House and Mathematical Association of America, 2004.

Enzensberger, Hans Magnus. *The Number Devil: A Mathematical Adventure.* New York: Metropolitan Books, 1997.

Herr, Ted, and Ken Johnson. *Problem Solving Strategies: Crossing the River with Dogs and Other Mathematical Adventures.* Berkeley, CA: Key Curriculum Press, 1994.

Kogelman, Stanley, and Barbara R. Heller. *The Only Math Book You'll Ever Need.* New York: HarperCollins, 1994.

Moretti, Gloria, et al. *The Problem Solver 6: Activities for Learning Problem-Solving Strategies.* Mountain View, CA: Creative Publications, 1987.

Polya, George. *How to Solve It,* 2nd ed. Princeton, NJ: Princeton University Press, 1988.

Rossi, Ernest Lawrence, Ph.D., with David Nimmons. *The 20 Minute Break.* New York: Jeremy P. Tarcher, 1991.

Chapter 10

Adair, Margo. *Working Inside Out: Tools for Change.* Berkeley, CA: Wingbow Press, 1984.

Beck, Aaron, M.D., and Gary Emery, Ph.D. *Anxiety Disorders and Phobias: A Cognitive Perspective.* New York: Basic Books, 1985.

Burns, David D., M.D. *Feeling Good: The New Mood Therapy.* New York: Avon, 1999.

Greenberger, Dennis, Ph.D., and Christine A. Padesky, Ph.D. *Mind over Mood.* New York: Guilford Press, 1995.

Gross, Ronald. *Peak Learning*. New York: Tarcher/Putnam, 1991.

Hackworth, Robert D. *Math Anxiety Reduction,* 2nd ed. Clearwater, FL: H&H Publishing, 1992.

Moses, Robert P., and C. E. Cobb. *Radical Equations: Civil Rights from Mississippi to the Algebra Project.* Boston: Beacon Press, 2001.

Pert, Candace, Ph.D. *Molecules of Emotion*. New York: Scribner, 1997.

Sapolsky, Robert, Ph.D. "Stress, Disease and Memory," Brain Expo 2000. Paradise Point Hotel, San Diego, CA. 18 January 2000.

Siegel, Daniel, M.D. "The Developing Mind," Brain Expo 2000. Paradise Point Hotel, San Diego, CA. 18 January 2000.

Chapter 11

Adair, Margo. *Working Inside Out: Tools for Change*. Berkeley, CA: Wingbow Press, 1984.

Beck, Aaron, M.D., and Gary Emery, Ph.D. *Anxiety Disorders and Phobias: A Cognitive Perspective.* New York: Basic Books, 1985.

Burns, David D., M.D. *Feeling Good: The New Mood Therapy*. New York: Avon, 1999.

Cameron, Julia. *The Artist's Way*. New York: Tarcher/Putnam, 1992.

Greenberger, Dennis, Ph.D., and Christine A. Padesky, Ph.D. *Mind over Mood*. New York: Guilford Press, 1995.

Gross, Ronald. *Peak Learning*. New York: Tarcher/Putnam, 1991.

Hackworth, Robert D. *Math Anxiety Reduction,* 2nd ed. Clearwater, FL: H&H Publishing, 1992.

Pert, Candace, Ph.D. *Molecules of Emotion*. New York: Scribner, 1997.

Sapolsky, Robert, Ph.D. "Stress, Disease and Memory," Brain Expo 2000. Paradise Point Hotel, San Diego, CA. 18 January 2000.

Siegel, Daniel, M.D. "The Developing Mind," Brain Expo 2000. Paradise Point Hotel, San Diego, CA. 18 January 2000.

Chapter 12

Afflack, Ruth. *Beyond Equals: To Encourage the Participation of Women in Mathematics*. Oakland, CA: Math/Science Network, 1982.

Berlinghoff, William P., and Fernando Q. Gouvea. *Math Through the Ages: A Gentle History for Teachers and Others*. Farmington, ME: Oxton House and Mathematical Association of America, 2004.

Drew, David. *Aptitude Revisited: Rethinking Math and Science Education for America's Next Century*. Baltimore, MD: Johns Hopkins University Press, 1996.

Answers to Master Math's Mysteries and Look Backs

(Chapter 1 Compatible Pairs: 75 + 25, 60 + 40, 35 + 65, 20 + 80, 95 + 5, 90 + 10, 375 + 625, 250 + 750, 100 + 900, 760 + 240, 450 + 550, 420 + 580)

Chapter 1 MMM

1. $2 + 450 + 550 = 2 + 1,000 = 1,002$
2. $6 + 1,250 + 750 + 3 = 9 + 2,000 = 2,009$
3. $300 + 75 + 500 + 25 = 800 + 100 = 900$
4. $3 + 900 + 100 + 10 = 13 + 1,000 = 1,013$
5. $6 + 400 + 100 + 20 = 26 + 500 = 526$
6. $1,000$
7. $7 + 590 + 1,410 = 7 + 2,000 = 2,007$
8. $175 + 325 + 5 = 500 + 5 = 505$
9. $25(3)(2) = 25(6) = 150$
10. 120
11. $5(3)(2) = 5(6) = 30$
12. 420
13. $(50 + 3) \cdot 2 = 100 + 6 = 106$
14. $4 \cdot (25 + 1) = 100 + 4 = 104$
15. $(10 + 1) \cdot 5 = 50 + 5 = 55$
16. $(10 + 2) \cdot 5 = 50 + 10 = 60$
17. $6(2 \cdot 5) = 6 \cdot 10 = 60$
18. $13 \cdot (2 \cdot 5) = 13 \cdot 10 = 130$
19. $(5 \cdot 2) \cdot 21 = 10 \cdot 21 = 210$
20. $42 \cdot (2 \cdot 5) = 42 \cdot 10 = 420$

Chapter 2 MMM

1. 1, 2, 3, 4, 5, 6, 7, 8, 9
2. 2, 4, 6, 8, 10, 12, 14, 16, 18, 20
3. 3, 6, 9, 12, 15, 18, 21, 24, 27, 30
4. 4, 8, 12, 16, 20, 24, 28, 32, 36, 40
5. 5, 10, 15, 20, 25, 30, 35, 40, 45, 50

Multiplication Quiz							
Row 1:	6	20	5	12	36	30	35
Row 2:	27	32	54	64	21	16	24
Row 3:	24	45	18	28	0	8	36
Row 4:	7	72	27	49	14	16	6
Row 5:	81	8	10	4	25	12	0
Row 6:	40	56	42	0	2	72	21

Chapter 3 MMM

1. $[7 + 18] - 14 = 25 - 14 = 11$
2. $5(2)^2 \div 2 = 5 \cdot 4 \div 2 = 20 \div 2 = 10$
3. $6 + 5(3) - 9 = 6 + 15 - 9 = 21 - 9 = 12$
4. $12 \div 3(3) + 9 = 4(3) + 9 = 12 + 9 = 21$
5. $12 - 3(3) + 9 = 12 - 9 + 9 = 3 + 9 = 12$
6. 11   7. 22   8. 4
9. 11   10. 36   11. 24
12. 6   13. 3   14. 21
15. 17   16. $\frac{3}{7}$ 17. 4
18. 4   19. 72   20. 7   21. 21

Chapter 3 Look Back

1. $[9 + 12] - 4 = 21 - 4 = 17$
2. $5(3)^2 - 2 = 5(9) - 2 = 45 - 2 = 43$
3. $12 + 3(6) - 9 = 12 + 18 - 9 = 30 - 9 = 21$
4. 5
5. 7
6. $1 + 625 + 375 + 4 = 1{,}000 + 5 = 1{,}005$
7. 500
8. $(50 + 3) \cdot 2 = 50 \cdot 2 + 3 \cdot 2 = 100 + 6 = 106$
9. $4(25 + 1) = 100 + 4 = 104$
10. $(10 + 1) \cdot 5 = 50 + 5 = 55$

Chapter 4 MMM

1. $\frac{1}{10}$ 2. $\frac{5}{37}$ 3. $\frac{5}{8}$ 4. $\frac{11}{12}$
5. $\frac{10}{x}$ 6. $\frac{5}{12}$ 7. $3\frac{1}{2}$ 8. $4\frac{1}{2}$
9. $1\frac{1}{2}$ 10. $1\frac{1}{2}$ 11. $7\frac{1}{2}$ 12. $2\frac{1}{2}$
13. $3\frac{1}{8}$ 14. $3\frac{3}{5}$

Chapter 5 MMM

1. 2, 4, 6, 8, 10   2. $\frac{4}{12} + \frac{3}{12} = \frac{7}{12}$ 3. $\frac{9}{12} + \frac{1}{12} = \frac{10}{12} \ or \ \frac{5}{6}$
4. $\frac{2}{12} + \frac{4}{12} = \frac{6}{12} \ or \ \frac{1}{2}$ 5. $\frac{2}{12} + \frac{3}{12} = \frac{5}{12}$ 6. $\frac{8}{12} - \frac{1}{12} = \frac{7}{12}$

7. $\dfrac{9}{12} - \dfrac{4}{12} = \dfrac{5}{12}$ 8. $\dfrac{10}{12} - \dfrac{3}{12} = \dfrac{7}{12}$ 9. $\dfrac{4}{12} - \dfrac{3}{12} = \dfrac{1}{12}$

10. $\dfrac{2}{12} + \dfrac{4}{12} + \dfrac{3}{12} = \dfrac{9}{12} \; or \; \dfrac{3}{4}$

Chapter 5 Look Back

1. $5 \cdot 2 \cdot 20 = 200$ 2. $3,000$ 3. $9 + 4 = 13$

4. $12 - 5 = 7$ 5. $\dfrac{4}{4} = 1$ 6. 106

7. 3 8. $5\dfrac{1}{2}$

9. $\dfrac{3}{7}$ 10. 350

Chapter 6 MMM

1. 359 2. 4.07 3. 0.204 4. 20.901

5. 4.0034 6. 14.8 7. 630.24 8. 0.0332

9. 335 10. 5.14 11. 0.398 12. 8.499

13. 3.9966 14. 3.6 15. 617.76 16. 0.0270

Chapter 7 MMM

1. $2 + 3 = 5$ 2. 6

3. 9 4. 7

5. $-4 + (-2) = -6$ 6. -8

7. -5 8. -7

9. -1 10. 0

11. 4 12. 1

13. -4 14. 2

15. $5 + (-4) = 1$

16. 6 dark beans $+ 3$ light beans $= -3$

17. -4 18. 2

19. 7 20. 3

21. -5 22. -4

23. 6 24. -2

25. 4 26. $-6 - (-2) = -4$

27. $-1 - 3 = -4$ 28. $3 - (-1) = 4$

29.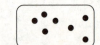

-5 $-$ $(+2)$ $=$ -7

30.

$$3 \qquad - \qquad (-3) \qquad = \qquad 6$$

31. 7   32. −6   33. 8   34. 6

35. −1   36. −9   37. 3   38. 9

39. −9   40. −2   41. −8   42. −6

43. −5   44. 5

Chapter 7 Look Back

1. −4   2. 2,000   3. $9 + (-4) = 5$

4. $-2 - 6 = -8$ 5. $\frac{7}{12}$ 6. 102

7. 15   8. $3\frac{1}{2}$ 9. 0.9   10. 530

Chapter 8 MMM

1. $\frac{1}{12}$ 2. $\frac{1}{6}$ 3. $\frac{1}{4}$ 4. $\frac{1}{6}$

5. $\frac{1}{12}$ 6. $\frac{1}{2}$ 7. $\frac{1}{12}$ 8. $\frac{5}{12}$

9. 80   10. −55   11. −18   12. −40

13. 1,000   14. −5

Chapter 9 MMM

1. 1   2. 3   3. 3   4. 10   5. 3

6. 4   7. 8   8. −11   9. −3   10. −30

11. 5   12. −20

Chapter 9 Look Back

1. 2.662   2. 20   3. 10

4. $\frac{9}{35}$ 5. $\frac{5}{12}$ 6. $\frac{7}{12}$

7. 8   8. 16   9. 3,000

10. 5.2   11. $\frac{13}{15}$ 12. 2

13. $\frac{3}{24}$ or $\frac{1}{8}$ 14. $\frac{5}{12}$

Chapter 10 MMM

1. 0   2. −2   3. 1

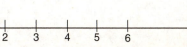

4. 1.5   5. −3.5   6. 0.5

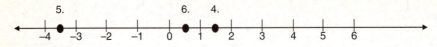

7. $-1\frac{1}{4}$ 8. $-3\frac{3}{4}$ 9. $4\frac{3}{4}$

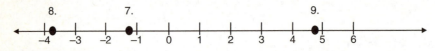

10. $-3\frac{1}{8}$ 11. 5.2   12. 0.7

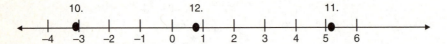

Chapter 11 Look Back

1. −2
2. 55
3. 5
4. −7
5. 9 · 4 = 36
6. 32 + 6 = 38
7. 9 + (−4) = 5
8. −2 − 6 = −8
9. 1
10. 5
11. 175
12. −4
13. 18 − 4 = 14
14. 25
15. $\frac{4}{6}$ or $\frac{2}{3}$
16. 450
17. 3.9
18. 21.137
19. $\frac{2}{15}$
20. $\frac{2}{12} + \frac{4}{12} + \frac{3}{12} = \frac{9}{12}$ or $\frac{3}{4}$

Chapter 12

1. some number plus 10 or 10 more than a number
2. six times a number minus two or two less than six times some number
3. five divided by a number or the quotient of 5 and x
4. four take away two is an unknown number or four minus two is some number
5. expression
6. expression
7. equation
8. expression
9. expression
10. equation
11. equation
12. expression
13. equation
14. expression
15. expression
16. equation
17. 2
18. 29
19. $x + 8$
20. $10 + 2x$
21. $x + 9$
22. 44
23. $9x$
24. $5x$
25. $7y$
26. $8y - 2x$
27. $7n$
28. $-3x$
29. 3 · 5
30. 2 · 5 · 5 or 2 · 5^2
31. 30
32. 4

33. −7   **34.** 14   **35.** 10   **36.** x is −8

37. x is 12   **38.** x is 26

39.

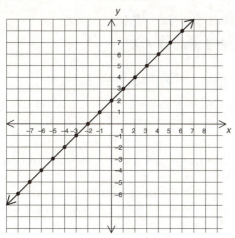

x	y
3	5
−1	1
0	2
−5	3

40.

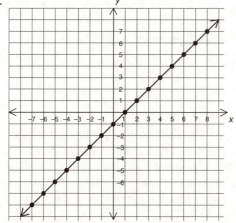

x	y
0	1
1	3
3	7
−2	−3

41.

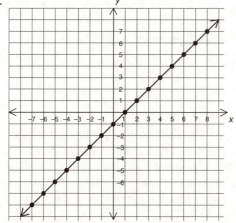

Additional Activities

Chapter 3—Assess Your Multiple Intelligences Activity

Model #1 Howard Gardner's Multiple Intelligences Model

Read these descriptions of each intelligence. As you read, consider how you would rate your skill in that area: strong, moderate, or weak. Shade the line from 1 to the number that you assess yourself for each intelligence.

1. Bodily-Kinesthetic Intelligence.   Ability to use the whole body or parts of it to develop products and solve problems as athletes, dancers, construction workers, actors, physical laborers, and surgeons do. (Famous examples: Roberto Clemente, Michael Douglas, Whoopi Goldberg, Karim Abdul-Jabbar, Michael Jordan, Shaquille O'Neal, Julia Roberts, Fernando Valenzuela, Serena Williams, Venus Williams, Tiger Woods)

1				5					10

Weak   Moderate   Strong

2. Interpersonal Intelligence.   Ability to understand motivations and inner workings of other people and to cooperatively lead or work with them. Teachers, mediators, negotiators, politicians, leaders, salespeople, and psychotherapists are examples. (Famous examples: Cesar Chavez, Henry Cisneros, Bill Gates, Mahatma Gandhi, Dolores Huerta, John F. Kennedy, Martin Luther King Jr., Golda Meir, Gloria Molina, Barack Obama, Eleanor Roosevelt, Mother Teresa)

1				5					10

Weak   Moderate   Strong

3. Intrapersonal Intelligence.   Ability in knowing and understanding one's own mental processes, effectively reflecting on thoughts, dreams, spiritual life, and motivations. Philosophers, authors, artists, psychotherapists, and many solitary individuals in all vocations exemplify this intelligence. (Famous examples: Stephen Aizenstat, Pierre Teilhard de Chardin, Milton Erickson, Karen Horney, Carl Jung, Thomas Merton, Claude Monet, Ernest Rossi)

1				5					10

Weak   Moderate   Strong

4. Logical-Mathematical Intelligence.   Ability to think mathematically and logically, as well as to analyze and reason scientifically. Accountants, inventors, mechanics, teachers, and engineers all put information together symbolically or practically using this type of intelligence. (Famous examples: Bhaskaracharya, Marie Curie, Thomas Edison, Albert Einstein, Lise Meitner, Isaac Newton, Emmy Noether, Ellen Swallow Richards, Julia Robinson, Chien-Shiung Wu)

1   5   10

Weak   Moderate   Strong

5. Musical Intelligence.   Ability in interpreting, performing, and composing music using melody, rhythm, and harmony. Composers, conductors, jazz musicians, music teachers, rappers, and cheerleaders exercise this ability as they create or use music in their work. (Famous examples: Desi Arnaz, the Beatles, Leonard Bernstein, Charo, Miles Davis, Julio Iglesias, Wolfgang Amadeus Mozart, Tito Puente, Poncho Sanchez, Barbra Streisand, Carrie Underwood, Ritchie Valens)

1   5   10

Weak   Moderate   Strong

6. Naturalist Intelligence.   Ability to see patterns and relationships in the natural world, classifying and discovering order. Scientists, biologists, botanists, and environmentalists exemplify this intelligence. (Famous examples: Rachel Carson, Jacques Cousteau, Charles Darwin, Albert Einstein, Rosalind Franklin, Barbara McClintock, Maria Mitchell)

1   5   10

Weak   Moderate   Strong

7. Spatial Intelligence.   Ability to form an abstract model in the mind of the three-dimensional world and then solve problems using that model. People who do this well include astronauts, sailors, muralists, engineers, architects, surgeons, graphic designers, sculptors, and painters. (Famous examples: Judith Baca, Franklin Chang-Díaz, Leonardo da Vinci, Sam Maloof, Michelangelo, Antonia Novello, Ellen Ochoa, Auguste Rodin, Helen Rodriguez, Vincent van Gogh, Frank Lloyd Wright)

1   5   10

Weak   Moderate   Strong

8. Verbal-Linguistic Intelligence.   Ability with words in writing, storytelling, discussing, interpreting, and talking. Poets, writers, lawyers, talk-show hosts, teachers, secretaries, and editors who form thoughts and use words skillfully in their work are examples of people strong in this intelligence. (Famous examples: Julia Alvarez, Maya Angelou, Agatha Christie, Emilio Estevez, Ernesto "Che" Guevara, Alex Haley, Langston Hughes, Edward Rivera, John Steinbeck, Victor Villaseñor, Alice Walker, Oprah Winfrey)

1   5   10

Weak   Moderate   Strong

Summarize Your Assessment

Record how you assessed your multiple intelligences. Shade each row across to the number you rated yourself.

Bodily-Kinesthetic									
Interpersonal									
Intrapersonal									
Logical-Mathematical									
Musical									
Naturalist									
Spatial									
Verbal-Linguistic									

1	5	10
Weak	Moderate	Strong

Write down your top three intelligences.

1. _____

2. _____

3. _____

Chapter 3—Assess and Understand Your Learning Modes Exercise

Model #2 You and Your Learning Modes

Quiz Yourself: Circle a, b, or c.

1. When you ask for directions, do you:
   a. Remember them verbally in your mind?
   b. Need a map or written instructions to follow?
   c. Move your arms and point to review the directions before driving off?

2. When you are a student in a classroom:
   a. Do you find it difficult to sit still and listen?
   b. Do you listen carefully and find noise distracting?
   c. Do you sit close where you can see what's going on and take notes?

3. When you are assembling furniture, do you:
   a. Move the pieces around and start putting them together immediately?
   b. Read the instructions and look over the diagram?
   c. Prefer to read the instructions aloud or have someone else read them to you?

4. Do you:
    a. Spell rather well and see words and pictures in your mind?
    b. Spell poorly but remember words from songs on the radio?
    c. Prefer activities in which you can move around and don't have to spell?

5. Would you be most likely to say:
    a. "I see what you mean."
    b. "I catch your drift."
    c. "I hear what you're saying."

6. Would you most likely use the phrase:
    a. "It slipped my mind."
    b. "I don't recall."
    c. "It appears I forgot."

7. For your birthday, would you most prefer:
    a. Lots of cards?
    b. Lots of phone messages?
    c. Lots of high fives?

8. Would you rather:
    a. Read a book?
    b. Go for a walk or a jog?
    c. Listen to the radio?

9. Would you rather:
    a. Cook a meal?
    b. Go to the symphony?
    c. Watch a movie?

Evaluate your quiz results by circling your answer to each question and then adding the circles in each column.

	I	II	III
1.	b	a	c
2.	c	b	a
3.	b	c	a
4.	a	b	c
5.	a	c	b
6.	c	b	a
7.	a	b	c
8.	a	c	b
9.	c	b	a
Score:			

Shade up to your score in the corresponding columns in the following diagram. The highest column *may* be your primary learning mode and the second highest, your secondary learning mode.

Continue to evaluate whether this assessment fits you as you read the descriptions and examples of each learning mode in Chapter 3.

	Visual I	Auditory II	Kinesthetic III
9			
8			
7			
6			
5			
4			
3			
2			
1			

Visual Learning

Visual learners require seeing. You form pictures and often see words spelled out or problems worked or situations happening in your mind. You may have a private movie screen in your head. When you understand something, you might say, "I see," and you do "see it" in your mind. You often forget verbal requests unless you write lists and reminders and put them where you can see them. You may speak slowly because of the difficulty of putting into words what you see in your mind.

Example of a visual learner: Sitting in front of the classroom, I see what's happening and am not distracted by watching other students. I take copious class notes, making little sketches to jog my memory. As a strong visual learner, I find that writing speeches down, recording them, condensing my speech to note cards, and then listening to my speech as I look at my notes or drive makes giving talks easier. Of course, I take my visual cues as backup. As a pianist since age six, I was "stuck" reading music—never playing without it—until I took a jazz piano class in which I learned to visualize chords and keys. Frequently I clear my mind for work by clearing my work space first to make it look neat.

Auditory Learning

Auditory learners rely on hearing. You listen to messages in your mind. You can repeat conversations or verbal input word for word. You often know all the words to songs you've heard. Radios, cassette players, MP3 players, and portable headsets play an important role in your life. The spoken word is essential. You may say, "I hear you" or "Sounds good" when you understand.

Example of an auditory learner: My artist friend, Emily, learns as she listens to tapes while painting and working. She is always conscious of her auditory background and travels with a cassette player and headphones. She tapes her textbook notes onto a cassette with her favorite music in the background. She often trades her artwork for someone else recording readings she wants to hear. Emily takes her recorder to the classroom and carefully chooses her location to avoid distractions from listening well. She can sing many '60s and '70s standards and has been known to recite 10-page poems "by ear."

Kinesthetic Learning

Kinesthetic learners need to move around and work manually with ideas. You touch things a lot. Smells and textures are important. You sometimes have difficulty sitting still in class just listening. The more activity you experience while doing a skill, the better you learn it. The more skin and muscles you use, the better you remember. Because most classrooms are taught by auditory or visual learners, you need to be creative to learn. Volunteer to move about the classroom acting as an assistant. Even small motions that seem unrelated to the activity such as swinging a leg, drawing, or knitting ease your integration of ideas.

Example of a kinesthetic learner: My social worker/teacher friend, Karin, moves through life touching and hugging. Interacting personally with her students, she keeps her adult classrooms active by using toys to illustrate ideas. As a student, Karin chose her teachers and mentors carefully so that her kinesthetic style was enhanced and supported. Her school projects involved activities to implement ideas. Cooking aromas, potpourris, and candle scents fill her comfortable home. The feel of her surroundings, not the look, is essential to her.

Chapter 4—Time Management Activity: Make Time for Math

Set aside all the time you need for math. Expect that math will be time intensive. Structure and control your time yourself. Don't let others control it. To build math into your schedule, do the following:

Step 1. Make multiple copies of the schedule.

Step 2. On the first copy, use pencil to plan your activities for the week ahead. Block off the hours for Sleep, Class, Study, Commute, Personal Care, Eating, Family Time, Housekeeping Chores, and so on. Build in flexibility for math study. Color-code the activities.

Step 3. At the end of the week, make a second chart with the activities that really occurred. Daily notes about your activities help you remember what you really did.

Step 4. Evaluate how you spent your time to discover whether you are spending your time on your priorities. Because passing math is a priority, schedule math study time.

Step 5. Plan your time for the next week and track yourself again. Continue to do this activity until your use of time fits your priorities.

	Monday	Tuesday	Wednesday	Thursday	Friday	Saturday	Sunday
6 A.M.							
7 A.M.							
8 A.M.							
9 A.M.							
10 A.M.							
11 A.M.							
NOON							
1 P.M.							
2 P.M.							
3 P.M.							
4 P.M.							
5 P.M.							
6 P.M.							
7 P.M.							
8 P.M.							
9 P.M.							
10 P.M.							
11 P.M.							

Chapter 9—Solutions for Stories to Ponder for Now

Story #5: Pets Solution

To see how many birds Claire had, we might remember that birds have 2 legs and the other animals (dogs and cats) have 4 legs. So, to find the number of birds she has, we can find how many animals have 2 legs. Also, we know that there are exactly 13 animals because each animal has only 1 head. We could work backward by assuming that Claire had 10 birds (a guess on my part) and 3 other 4-legged animals. Counting 2 legs for each bird and 4 legs for each of the others, we find that our assumption (10 birds) gives us $2 \cdot 10 + 4 \cdot 3 = 32$ legs. By the way, this method is Guess and Check. Our assumed answer was a guess, and our check showed that we were incorrect. Our assumed answer gave us 32 legs—not 36 legs like we need. This tells us that we need more legs. Now, you try it. Make another assumption about how many birds Claire has and check that assumption with the facts: 13 heads and 36 legs. Another way to work this problem would be to draw pictures.

Story #6: The Diaz House Solution

To see if Mrs. Diaz can afford the molding, we need to know her cost and compare that to the budget of $200. To find her cost, we need to know how many yards the job requires. The room measurements are in feet, so a drawing of the room will be a start. See the room diagram with the work.

13	3 feet = 1 yard
11	48 feet = 16 yards
13	
+11	
48 feet around	

Cost is 16 yards • $10 = $160.

Because the crown molding cost is $160, which is less than her $200 budget, Mrs. Diaz can afford it.

Story #7: Mowing the Lawn Solution

Amy must have thought fast. Maybe she used her fingers to keep track of the days, doubling the money as she moved across her hands. Try it. One finger = 1 penny. Two fingers = 2 pennies. Three fingers = 4 pennies. Four fingers = 8 pennies. Five fingers = 16 pennies. Six fingers = 32 pennies. Seven fingers = 64 pennies. Eight fingers = 128 pennies. On the eighth day she finally gets more than a dollar. Nine fingers = 256 pennies. Ten fingers = 512 pennies or $5.12.

Let's round $5.12 down to $5 and keep going with the fingers. Eleven fingers = $10. Twelve fingers = $20. Thirteen is $40. Fourteen is $80. Fifteen is $160. Wow! If Amy can wait until Day 15, she receives over $100 in just one day. And, after Day 15, the money keeps on doubling!

Amy gets very curious about how much money she'll receive on Day 30, so she says good-bye to her friends and walks home for a pencil, paper, and calculator. Recording each day on a t-table and beginning again, Amy notices that the amounts of money are always powers of two. That means that each amount of money can be calculated by multiplying twos together. Look at Amy's t-table. Day 2 is 2 cents. Day 3 is 2 • 2 or 4 cents. Day 4 is 2 • 2 • 2 or 8 cents. And so on. Amy notices a pattern. She sees that the number of twos is just one less than the number of the day. She continues the t-table forward to check out this pattern. Yes, Day 5 is four twos multiplied by each other to get $.16. Day 6 is five twos multiplied by each other to get $.32.

Day Number	$	
1	1 penny	
2	$2 = 2 = 2^1$ ◄————	Amy sees that
3	$4 = 2 \cdot 2 = 2^2$ ◄————	this exponent
4	$8 = 2 \cdot 2 \cdot 2 = 2^3$ ◄————	is one less
5	$16 = 2 \cdot 2 \cdot 2 \cdot 2 = 2^4$ ◄————	than the "day
6	$32 = 2 \cdot 2 \cdot 2 \cdot 2 = 2^5$ ◄————	number."
7	$64 =$	
8	————	
9	————	
.		
.		
.		
30	————	

Because Amy knows about exponents and her calculator, she knows that $2 \cdot 2 \cdot 2 \cdot 2$ is written 2^4 and she can use the $\wedge$ or y^x calculator key to verify this. (Calculators differ, so check your manual.) Amy checks her calculator technique by checking that $2^5 = 32$ and $2^6 = 64$ before she even tries to calculate 2^{29}, the amount of money that her uncle would give her on Day 30. When she is sure she knows how to calculate powers of two, Amy takes a deep breath and punches in 2^{29}. She takes a look at the answer and screams, "I'm rich." How much money will she receive on Day 30? Would you have made the same choice as Amy? Can Amy's uncle afford it? Did Amy mow his lawn on April 1? What happens on Day 30?

Story #8: Baking Cookies Solution

Find the total number of cookies that Emilio baked. Begin at the end, when Emilio has six cookies, which make up the half that Emilio did not eat. Emilio must have eaten six cookies. One half is the six remaining, and the other half is the six he ate. Those are the 12 cookies that he did not give Julia. In fact, they are the half she didn't get, so she got 12 cookies. One half is Julia's 12 cookies, and the other half is Emilio's six and the remaining six. Look at this diagram to see what is happening.

		plus one dozen burned cookies.
6 left.	Emilio ate 6.	
Julia gets 12.		6 left
Put away 24 for younger brothers.		+ 6 for Emilio

6 left
+ 6 for Emilio
+ 12 for Julia
+ 24 for brothers
+ 12 burned
60 cookies

Continuing back to the beginning, Julia's 12, Emilio's six, and the remaining six cookies (or 24 total) make up the half that were not put away for the younger brothers. That means the brothers get 24 cookies. This time, one half is the 24 for the brothers and the other half is eventually shared (unevenly) by Julia, Emilio, and the leftover plate. Because Emilio burned 12 cookies, the total number of cookies that he baked is 60.

Problem-Solving Index

I N D E X